PENGUIN BOOKS

# JOY, GUILT, ANGER, LOVE

GIOVANNI FRAZZETTO was born and grew up in the southeast of Sicily. In 1995, after high school, he moved to the UK to study science at University College London, and in 2002 he received a PhD from the European Molecular Biology Laboratory in Heidelberg, Germany. Since he was a student, he has worked and written on the relationship between science, society, and culture, publishing in journals such as *Nature*. He now lives and works between London and Berlin.

# Joy, Guilt, Anger, Love

## What Neuroscience Can—and Can't—Tell Us About How We Feel

GIOVANNI FRAZZETTO

PENGUIN BOOKS

PENGUIN BOOKS
Published by the Penguin Group
Penguin Group (USA) LLC
375 Hudson Street
New York, New York 10014

USA | Canada | UK | Ireland | Australia | New Zealand | India | South Africa | China
penguin.com
A Penguin Random House Company

First published in Great Britain as *How We Feel* by Doubleday, an imprint of Transworld Publishers, 2013
Published in Penguin Books (USA) 2014



ISBN 978-0-14-312309-5

Printed in the United States of America
10 9 8 7 6 5 4 3 2 1

Set in Sabon
Designed by Spring Hoteling

*To Berlin's river tour company Reederei Riedel*
*and in memory of*
*Yehuda Elkana*
(1934–2012)

# Contents

# Prologue

When I worked in a neuroscience laboratory, the rhythm of experiments paced the hours. The lab was an island, a hideaway that felt distant from reality. It was a world of its own, one I had desired to set foot in ever since I turned sixteen. Inside, there was always quite a lot to do: exact solutions to prepare, delicate dissections to perform, precious molecules to purify and animals to take care of. Tightly parcelled one after the other as in a chain, these were some of the tasks that punctuated the flow of my daydreaming and at the same time pointed to big research questions. In between, I filled my lab journal with notes, diagrams and calculations. Trying to understand something as ineffable and intimate as emotions and the mind, I amassed minute fragments and discrete units of technical information.

Venturing into the secrets of the human brain became an opportunity for deep reflection. It was like interrogating an unfamiliar aspect of myself. It was like deciphering a tale written in code about the mind that I myself, with my experiments, was contributing to writing. Brain tissues, neurons and stretches of DNA were the protagonists of a story that, fact after fact, revealed new truths.

Every evening, with my lab coat dirty, my lab journal stained with chemicals, and standing in front of empty glassware piled up in the sink, I would assess the progress I had made. Usually my thoughts were also in need of a rinse. No matter how much I had laboured at the bench, there always seemed to be something left undone. One question demanded another, every experiment begged for confirmation, the results could use a second round of analysis. But the next chapter of the story was always scheduled for the following day.

When I made my way home, the characters from the lab would stay behind and I would latch on to another story still in progress, that of my own emotional life, of which I was the only protagonist, with my own script, the lines and movements of which were also still to be discovered. At home, I was face to face with my emotions.

Emotions, even the most fleeting, pervade every portion of our lives. One minute we are sad, the next we are beaming with hope. Some emotions chase us, others elude us. Every so often, emotions may leave us wounded, or they may consume us. On other occasions they lift us or transport us afar. This is why, sometimes, we think it would be useful to know how to rid ourselves of some of our emotions, or at least learn how to tame them. Occasionally, as in the case of joyful emotions, we wish we could make them recur on demand.

While I was writing this book, whenever I revealed to new acquaintances that I work as a neuroscientist, they, no matter their field, would want to know more. If I then mentioned emotions, there was no risk of failing to strike up a conversation. I found that people would ask me for advice on how to control their temper, how to forget unpleasant memories, how to overcome fears and cultivate joy, and even how to fix or save their love relationships. And they were unfailingly surprised when, *even though I studied the brain*, I didn't always have answers for them.

We have it from the ancient wisdom of Socrates, the great Athenian philosopher, that discovering the exact causes of a phenomenon does not concurrently reveal its meaning for us and our lives. It seems that in the last days before his death, around 399 BC, Socrates read a book by Anaxagoras, a leading contemporary scientist. He had heard the news that Anaxagoras had discovered an element called *nous* (mind) that explained the nature of all things.[1] Socrates hoped to learn the riddles of existence with the help of that book. However, when he realized that *nous* was only a force that ordered nature's elements – air, for instance, or water – and could not tell him much about the meaning of life, let alone how it should be lived, he was filled with disappointment. Science was no road to self-knowledge.

This question – how to harvest scientific knowledge so as to learn how to live, or to know oneself, for that matter – grew no less urgent in the millennia to come. At the end of graduate school I came across a revealing essay: the transcript of a lecture delivered in 1918 by the German sociologist and philosopher Max Weber (1864–1920) and entitled '*Science as a Vocation*'.[2] Going by its title, I was hoping to find there an echo of my passion for research. In the essay Weber addresses an audience of young students on the meaning and value of science for both personal and broader questions in life. Its take-home message was not encouraging. For Weber, science was responsible for a process of profound intellectual rationalization, which he termed a disenchantment – in German, *Entzauberung*. Science meant human progress, yes, but it was not necessarily synonymous with a life full of existential meaning, because science teaches us only how to master life 'by means of calculation'. I had a strong reaction to that essay. How could science ever be meaningless, or of no value?

My wonder at science remained unscathed, but Weber's question about how it could help me understand life, or myself, resonated loudly.

In fact, almost a century later, that question grows ever more pressing for us. At the dawn of the second millennium, we live in a world that is profoundly pervaded by science and technology. The incredible amount of information about the brain at our disposal delivers the resounding message that what counts most in us is a web of neurons and that, if we learn how those neurons work, we will come closer to understanding who we really are. An enthusiastic belief reverberates: deciphering the mysterious code of the brain would let us adhere to the ancient dictum 'Know Thyself', proving Socrates wrong by successfully using science to throw light on our existence – even in that most private and shadowy territory, our emotions.

But can the neural script of the brain indeed tell us how we feel?

This book unfolds as a collection of stories that contribute to answering that question. While providing a version of what neuroscience has unravelled about our emotions, I shall also tell you what such discoveries have meant to me as I studied the brain and walked the spine of life. Chapter by chapter, I will disclose when the neural subtext to the emotions I experienced clarified and embellished some of their qualities, but also when it remained a mere appendage to what I felt. Episodes of anger, guilt, fear, sadness, joy and love will reveal how the neural tapestry of an emotion can be an endless source of wonder, but also leave us with knots to untangle.

# 1

# Anger: Hot Eruptions

Anger dwells only in the bosom of fools

ALBERT EINSTEIN

Anybody can become angry – that is easy, but to be angry with the right person and to the right degree and at the right time and for the right purpose, and in the right way – that is not within everybody's power and is not easy

ARISTOTLE

It was one of those mornings when you know the day is going to turn ugly and everything will go wrong.

You know it already, because you've been kept awake all night by a neighbour's barking dog, mosquitoes have somehow managed to seep through the window nets and, when you've finally succeeded in falling asleep, a misdialled call wakes you at the crack of dawn. And if all that weren't enough, once up and on your feet, you spill hot coffee over yourself. Yet you have no alternative. You need to start your day, accept that life is hard and venture into the unknown, come what may. In fact, what lay ahead was not at all a terrible prospect. I was taking a holiday in Rome and friends had kindly organized a one-day trip to their country home, not far from the centre of town, to chill out and spend a long afternoon together.

'Bruce is going to pick you up,' they told me.

I sipped an espresso and waited for the stranger outside the hotel. It was hot and temperatures were sure to rise further.

'Nice to meet you, Bruce!' I said openly when his car stopped in front of me. 'How is it going?'

'We don't have much time, there is a lot of traffic in town and we need to hurry up. Get in the car, quick!'

Right, someone else has had a rough night, I thought, and I obeyed the order, already looking forward to reaching our destination.

The roads were indeed packed and everyone seemed in a rush. It took us an hour to leave the centre, zigzagging between erratic cars and through a swarm of revving mopeds coming from all directions. A few minutes into the journey I had a clear lesson in physics. I came to realize that in Rome the real duration of the infinitesimal moment, the shortest period of time, is the span between the green light and the person in the car behind you pressing the horn and yelling. In the meantime, Bruce wouldn't stop complaining about everything: all other drivers were either too slow or too fast, idiots and jerks. As we finally reached the motorway, our air-conditioning system broke down and ahead of us was an endless sea of cars.

'Great, what else now?' Bruce barked.

Things certainly didn't look promising, I thought, but pulled down the car window and leaned back, yielding to the helplessness of the situation.

But Bruce didn't relax. He began tapping his fingers on the steering wheel. Each tap a unit in a pounding countdown, like the last drops of his ebbing patience.

'Everything all right, Bruce?' I dared to ask.

He didn't really pay attention to what I said, his eyes fixed on the cars stretching ahead. He started to honk aggressively, pulling down his window. A series of expletives were hurled at the other cars, as if the honking and his verbal eruption could together make the queue recede.

Half an hour later, when miraculously we reached our exit,

someone who had cheekily been driving along the emergency lane emerged from one side, cut us off . . . and gave us the finger!

That's when I began to be afraid for everyone's safety. The sky outside was limpid, but Bruce's face looked like a thunderstorm. It was as if he felt trapped in a narrow corridor and his only goal was to get out of it as fast as possible. He just jumped out of the car and started to yell at the other driver, who, meanwhile, had quickly vanished. Everyone behind began protesting that we were obstructing the exit and Bruce raged at them all, telling them to shut up. Luckily, no one got out of their car and, before he could run towards them, I grabbed him and pulled him back inside.

We got out of there and, to be honest, all I could think about was a hammock under the shadow of a tree at my friends' house.

## An overwhelming force

Anger is a crude emotion, a mighty force that can be very difficult to contain. For it to surface, it may be enough that events simply run contrary to the way we expect them to go for us. We express anger if we are ill treated, if we feel we have been slighted, if someone offends us or when we won't or can't tolerate certain kinds of behaviour. Anger is also fear with an armour. It works as a defensive, pre-emptive reaction before something hurtful can be done to us. Anger may be impulsive and spontaneous, acted out impetuously in brief and acute bursts, but it may also be silent and premeditated, lucid and controlled. It can be both an immediate response to provocation and the fuel for future retaliation. What is interesting about anger is that it can lurk restrained for a long time, erupt wildly and fleetingly, and then return to a quieter state. Once past the hot, flashing eruption of a blinding fury, you can remain angry at someone for a long time. In all its forms, anger inescapably entails morality. The inability to control impulsive reactions puts our character to the test, and can be seen as weakness or a defeat of the will. Yielding to rage may have repercussions for our position in the social world and may jeopardize our interpersonal relations.

Of all emotions, anger is, for sure, the one that is most foreign to me. I am not irascible and I am not given to rages either. I can engage in a short verbal argument to resolutely put my point across because I dislike being misunderstood and disregarded in a conversation. On occasion I have also incisively pointed out my rights during over-heated conversations with customer services hotlines. But I never get into an aggressive verbal fight or, far less, become physically aggressive or abusive. I was never attracted to violence. However, one situation where I believe I could express extreme anger would be if anybody deliberately harmed a member of my family, or one of my best friends, especially if they did it before my eyes.

So why did Bruce react so vehemently to an unexpected Saturday morning queue? Why couldn't he deal better with his frustration, and what pushed him to yell at the other drivers?

When we reached our friends' cottage, Bruce wound down with an ice-cold drink. During a quiet moment he started telling me he had experienced similar outbursts of anger in the past. In certain situations he turned ugly, then later regretted it and wasn't happy about it. Especially when provoked, he was often incapable of controlling his reactions and this, of course, worried him. If someone contradicted him or didn't agree with him, he would make a fuss and, occasionally, put up a fight. But his rages could also take place in solitude. Once, angry at a small offence received at work, he smashed his own car's windscreen as a way to vent his frustration. He thought there was something wrong with him and asked me if his repetitive and uncontrollable explosions of anger might have something to do with his genes and the hard-wiring of his brain.

Clearly, some of us are more prone to anger than others. Why is that? Are we born aggressive, or is the propensity to express anger a consequence of upbringing, or a response to socially negative experiences or an unfavourable environment?

In this chapter I am going to address this question by telling you what we know about the neuroscience of anger and violence and what brain mechanisms underlie self-control.

But first, there is a lot I need to tell you about emotions in general.

## The origin of emotions

It would be unthinkable to talk about emotions without evoking the work of Charles Darwin. The brilliant British naturalist, most famous for having fathered the theory of natural selection and evolution, did not overlook the importance of understanding how we emote. In 1872, about a dozen years after *On the Origin of Species*, Darwin published a beautiful volume called *The Expression of the Emotions in Man and Animals*, his biggest legacy to the field of psychology.[1]

Darwin based his work on a few original resources. First, during several dinner parties at his country home in Kent, Darwin asked his guests to describe and comment on the emotions they recognized in a series of pictures. The pictures he showed them were eleven black-and-white photographs taken by the French anatomist Guillaume-Benjamin-Amand Duchenne. These depicted an elderly man's face, to which Duchenne had applied galvanizing electrodes to specific muscles to trigger a variety of facial expressions. Darwin asked his guests to describe what emotion they thought the man's face showed. An indefatigable collector, Darwin had never ceased to look for portrayals of emotion. He scoured galleries and bookshops for images and prints that could further his research. Eventually he also teamed up with a photographer, Oscar Rejlander, to help him capture fleeting moments of the emotions he was looking for. Although Darwin's experiment is not considered scientific by modern standards – for he relied on only twenty-three guests and his sources were diverse and of debatable objectivity – it was an extremely original and revealing enterprise for that time. Darwin's use of photographs and portraits also marked a huge leap in the history of scientific illustration.[2]

The main merit of Darwin's book was that it portrayed emotions as an outcome of evolution. By describing in detail emotional expressions in animals and human beings, Darwin made the point that emotions are comparable across the animal kingdom. By this he didn't mean that, say, the rage experienced by a human can be fully equated to the angry barking of a dog, or that human anxiety is exactly the same as a cat's fear, but that the evolutionary purposes of the mechanisms of defence and protection behind these emotions are analogous. Darwin meant that each emotion has adaptive purposes and has its evolutionary origins in lower animals. Just like our eyes, legs or other parts of our anatomy, emotions – and all the brain circuits and body parts that we need in order to experience them – have also evolved by natural selection. Within this general framework, it becomes easy to appreciate that the importance of Darwin's penetrating survey lies in its confirmation that emotions are first and foremost something that happens to the body: a physiological response to the events in the environment – or, of course, a consequence of thoughts and imagination recalling them – that is manifested through various physiological changes.

This view essentially persists today in light of modern neuroscience research and research into emotions in lower animals, such as rodents. Most people ask with scepticism: how can you study anger, joy or anxiety in a mouse or a rat? The answer is simple: you can't. What is explored in the laboratory are only the universal aspects of emotion, those accomplished by dedicated circuits that allow animals and human beings to survive and thrive.[3] In evolutionary terms, Darwin's study of expressions suggested that all organisms display innate and conserved primordial emotional mechanisms that help them survive. At opposed extremes on a gradient of such mechanisms are *approach* and *avoidance*, which are strategies for, respectively, achieving pleasure and shunning pain. For instance, available food and sex are clearly powerful motivators for approach because they bring joy and gratification – in addition to promoting survival and reproduction. By contrast, predators or other dangerous situations that cause fear prompt escape and evasion. In order to survive, we must be able to experience both approach and avoidance. These two principal survival mechanisms have been maintained throughout evolution and are shared across the animal kingdom and across different human cultures. With joy and fear at its opposite ends, there is an emotional rainbow of positive and negative emotions. The distinction here is not between good and bad. Again, a good guiding principle is that of approach and avoidance. The negative emotions are anger, guilt, shame, regret, fear and grief, all of which imply something we need to defend ourselves from or avoid. The positive emotions are empathy, joy, laughter, curiosity and hope, which all imply a propensity and desire to open up to the outside world.

At this point there is another important distinction to be made: between emotions and feelings. Feeling is emotion which has been rendered conscious. Although emotions develop as biological processes, they culminate as personal mental experiences. The contrast here is between the outer and visible aspects of an emotion and its inner, intimate experience. The former is a collection of biological responses – from alterations in behaviour and hormonal levels to changes in facial expression – that can, in most cases, be scientifically measured. The latter is the *feeling*, the private awareness of that emotion (philosophers call the study of this subjective experience phenomenology).[4] This is why we can describe our own feelings fairly confidently but we can't describe the internal experiences of others with the same degree of confidence. We can only watch their outward expression and theorize or intuit the inner experience of others. So far, in a laboratory, scientists can detect some of the brain activities that characterize sadness or joy. Yet they can't grasp the most internal *meaning* of sadness or joy for the person who experiences it. Emotions make our minds speak to each other. They are the most faithful reproduction of our inner worlds, broadcast externally in the expression of our faces.

Darwin's second important achievement in the study of emotions was his demonstration of their universality. If emotions are innate and a product of evolution, he hypothesized, they should also be widespread and similar across cultures. If all humans around the globe possess the same eyes, mouth, nose and facial muscles, then they should all be equipped to manifest emotions similarly. To show this, he adopted the methods of the anthropologist. He sent a detailed set of questions on all kinds of emotions to cultivated friends and other scholars, as well as to missionaries who travelled in then remote lands such as Australia, New Zealand, Malaysia, Borneo, India and Ceylon. He received thirty-six answers. This was probably one of the first printed surveys ever produced. Darwin asked his correspondents to report whether populations in those distant cultures, and in particular aboriginal tribes, displayed facial expressions and bodily postures comparable to those he was familiar with in Britain and Europe.

Darwin's oeuvre is an absolute treasure for the understanding of emotions. It has left a lasting legacy and has inspired many other scholars in this field.[5] I will make repeated reference to Darwin when describing the main bodily, and particularly facial, features of emotions. Let's begin by looking at the facial features of anger.

## Anger's ugly face

Not only was Darwin an incredibly original thinker: he was also a clear and expressive writer. His descriptions are so concrete and accurate that, even when he has no photographs to show, you can visualize the bodily changes he is writing about.

In the case of anger, Darwin remarks that the 'heart and circulation are always affected'. Indeed, there is nothing like a fit of anger to get your blood flowing and bring on a sudden hot rush – try it, especially if you are feeling cold. Your veins fill up with blood and distend, becoming prominent, especially on your forehead and neck. Blood flows into your hands, as if to prepare them for defensive action. Darwin knew that the arousal of anger involved the brain, and he makes this explicit when he says that the 'excited brain' sends vigour to the muscles and 'energy to the will'.

All in all, anger is an electrifying emotion. It empowers us to take action. An angry face 'reddens or becomes purple'. In anger, we glare. Darwin also notes that, in anger, the mouth commonly stays firmly closed to convey determination, and the teeth usually grit. Occasionally, however, the lips may retract to uncover the teeth, as if to show defiance to those who offend us.

Anger also alters the voice. During an explosion of angry speech there may be so much ferment and uproar that the mouth 'froths', as Darwin put it, and words become confused. Indeed, when unbridled, anger is most often a loud emotion, discharged as strident, rowdy, rapid sounds. One thing is certain about anger: it escalates. And you can see it mount in the face of a person who is in a rage. Not only that: it's as if the whole body heaves and swells up until it finally explodes into verbal and physical outbursts.

## An unjustified war

Anger exemplifies the irrepressible vigour of the emotions. It puts our judgement to the test, forcing us to consider how to behave in frustrating circumstances, respond adequately to offence, and decide on the best action. Anger is entangled with choice. Feeling anger raises questions of values and options, and thus of ethics, morality and conduct.

For too long in the history of ideas, when it came to finding an explanation for how we exercise our judgement, an over-rigid and simplistic assumption held sway. This categorically divorced emotion from reason, seeing the two as opposing poles in our mental life. Morality was held to be firmly grounded in logical reasoning, while emotion had nothing to do with it. This divisive theory, until recently so engrained in our culture, originated more than two thousand years ago, in ancient Greece, the cradle of Western thought, mainly in the writings of the philosopher Plato (427–347 BC), a diligent pupil of Socrates.

Plato's thoughts on emotion and reason are most explicitly laid out in the *Republic*, his essay on morality and the ideal state, but also in its sequel, the *Timaeus*, where he sketches out his ideas on the physiology of the soul and makes clear reference to the parts of the body that he believed hosted it.[6] According to Plato, the human soul was animated by three main types of passions or energies: reason, emotion and the appetites. Of the three, reason was by far the noblest, whereas emotion, and even more so the appetites, were second-order passions, granted lower status. The appetites were our basic needs, such as those for food and sex, as well as greed for money and possessions. Emotion was impulsive, unguarded reactions, such as anger or disgust, but also bravery. By contrast, reason meant calm reflection, zeal, persuasion and argument. Conveniently, the tripartite soul mirrored Plato's triune social division of the state. The lowest class, the proletariat, embodied the appetites, notably stinginess and greed. Among the warrior class the emotions dominated. The guardians, the highest class in Plato's society, personified reason.

It was to reason that Plato granted most importance, claiming that only a rational man could be both just and moral. Basically, the passions must submit to whatever reason dictates. This tripartite idea of the mind flourished, in different permutations, almost unquestioned for about two millennia.

Sigmund Freud (1856–1939), a Viennese physician and the father of psychoanalysis, who certainly believed in the importance of emotions, parcelled rationality off from basic instincts and acknowledged the conflict between the two. The most primitive human desires constituted what Freud called the id, Latin for 'it'. This vague, amorphous part of the human mind encapsulated its most visceral instincts. The workings of the id are free of all logic or rationality. They are also outside our conscious perception and control. Essentially, the id is the most rudimentary mental survival mechanism, the one we share with all lower animals and that we are born with, and it has two main objectives: the attainment of pleasure and the avoidance of pain (Fig. 1).

Above the id in Freud's hierarchy of the mind was the ego, which constituted human rationality. The ego works both consciously and unconsciously. In its conscious form, the ego is what takes care of the mind's perception of and relationship to the outside

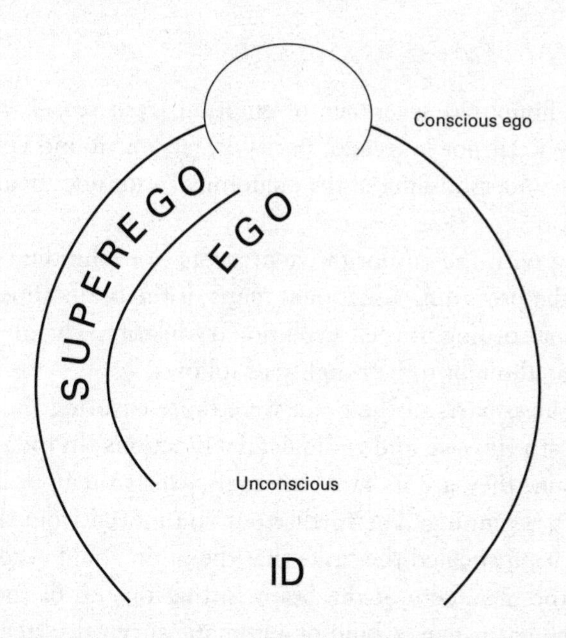


Fig. 1 Freud's structure of the human mind. Most of our mental processes are unconscious, floating underneath our awareness. Only a tiny part of our thoughts and emotions is fully conscious (top of diagram). The id represents our most visceral instincts. The ego is the seat of our rationality and consciously governs our relationship to the outside world. It also unconsciously represses some of the id's instincts. The superego represents our sense of morality, shaped by society and culture. (After a diagram in *New Introductory Lectures on Psychoanalysis*, 1933, Lecture XXXI)[7]

world, through the five senses. The ego is what makes us plan ahead. Through its unconscious qualities, the ego also exerts inhibitory control on the id, repressing some of its instinctual drives.

Lastly, at the top of the ladder, was the superego, our conscience and the repository for our sense of guilt, a moral apparatus moulded by society and culture.

Despite Freud's initial interest in the brain – he started his career as a highly regarded neurologist – the physical location of these components of the human mind did not concern him. Even so, he remarked several times that his psychological theory of mind would one day be replaced by a physiological and chemical one. And his prediction would be confirmed.

## Ideas cemented in the brain

The time-honoured severance of emotion from reason retained credibility until not long ago, partly because it found confirmation in the understanding of the anatomical and functional design of the brain.[8]

Fitting with the authoritative precepts of evolutionary development, the prevailing functional maps of the brain allocated its functions according to their evolutionary history. The division of labour was thought to be roughly as follows.

The oldest parts of the brain were those ensuring the control of the most primitive and rudimentary functions. In today's map of the brain, they are its most internal parts: in temporal terms, the brain's beginning. The further out you moved from the core, the more sophisticated the tasks that the brain could accomplish. Deep in the meanders of the brain, sitting on top of the spinal cord, is the brainstem, a kind of automatic survival system, without which we wouldn't even be able to breathe. The brainstem is the pillar of our physiological existence (Fig. 2). It contains structures such as the medulla, which controls breathing and heart rate, and sends and receives signals to and from vital organs. It can be thought of as the general 'power switch' of the brain. If something happens to the brainstem, the whole system shuts off, and that is why, for instance, an injury to the brainstem in a fall or other kinds of accident is fatal. Closer to the outside surface, but still in the core of the brain, are units whose function adds emotions to the basic survival mechanisms of the brainstem. It is within these deep structures that emotions in their rawest form are processed. Together these structures – which, roughly speaking, include tissues with extravagant, almost mythological-sounding names such as the thalamus, the hippocampus and the amygdala – are called the limbic system. 'Limbic' derives from the Latin *limbus*, which means border or edge, an appropriate name for this set of tissues which protrude from and cover the brainstem.

Lastly, wrapped around the limbic system and the brainstem, is the cortex (Latin for bark). The cortex is the last addition to the

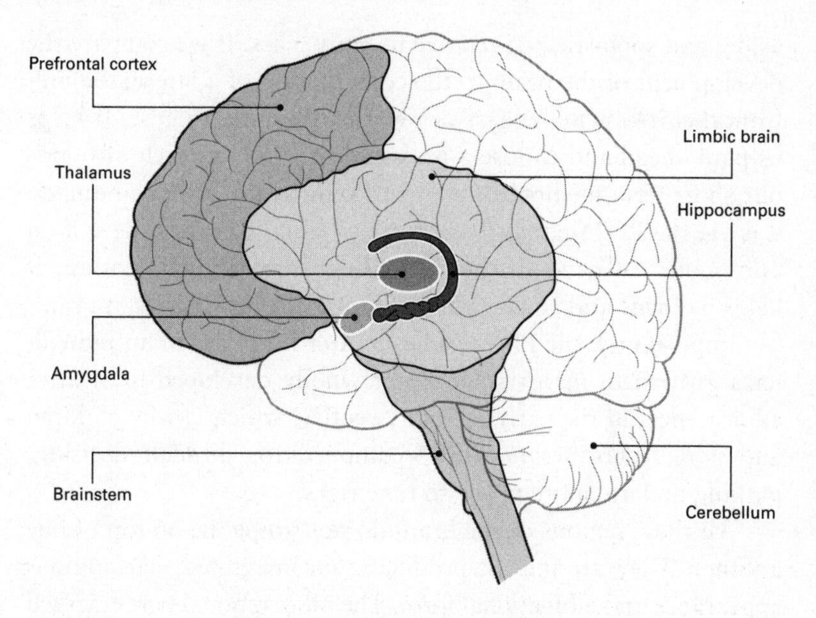


Fig. 2 Schematic diagram of the brainstem, limbic system and neocortex

brain yet it is the most evolved. When the cortex first appeared, it was rather thin. Over time – evolutionary time, therefore millions of years since it first appeared – it continued to grow within the boundaries of the skull, increasing the number of neuronal cells and therefore its capacity. In mammals, over a hundred million years ago the cortex underwent remarkable growth and became what is now called the neocortex, the most sophisticated version of the cortex.[9] Covering the rest of the brain like a cap, it is made up of large, convoluted folds of tissue that make the brain appear like a wrinkled sheet.

The areas where the tissue turns are called gyrii, while the intervening furrows are called sulci. Within the neocortex, the part that has undergone most of the change and that has grown in mammalian history is the most anterior part, the prefrontal cortex, or PFC, just behind your forehead and eyes.

The PFC occupies almost one third of the entire volume of the cortex. We are the only species on earth with a prefrontal cortex as big and sophisticated relative to body mass. If we compare the development of the brain to the construction of a house, the prefrontal cortex is its highest storey, the brain's lofty attic. It helps us plan ahead and choose a preferred path of action. It also aids our short-term memory. If someone tells us their phone number, it is via the PFC that we keep it in our mind before we save it on our phone. In general, the PFC also controls our attention. It helps us focus and concentrate, and not drift away from a task.

Importantly, the PFC reaches its full form late in an individual's growth to maturity. It is not wholly developed until after adolescence, in the early or mid twenties, which is why children and adolescents are not fully equipped for difficult decision-making and are more prone to take risks.

All these regions of the brain do not simply lie on top of one another. They are joined, producing an integrated, harmonious appearance and a functional form. The more rational part emerged from the existing impulsive core and, as a result, the two are densely and strategically connected to communicate with each other and regulate emotion.

So, for many centuries, rationality and emotionality were considered two opposing properties of the brain, operating as competing territories. They were like two substances that repelled each other and never mixed, rather like oil and water. The rational brain helps us analyse facts and assess external events, while the emotional brain tells us about our internal states.[10] During the past two decades this rough division of labour in the brain has been challenged. The brain's geographical boundaries as regards the accomplishment of rational tasks and emotion have blurred. The prefrontal part of the brain still holds the reins of rationality, but it also contributes to emotion.

This crucial and fascinating reversal in the understanding of the role of emotion has been underpinned by experimental work, particularly that of neuroscientist Antonio Damasio. Before we consider this, I need to tell you a story.

## Skin reactions

It is common practice in biology and medicine to understand the ordinary mechanism of the function of a tissue, an organ or even a gene by simply observing what happens when that function is removed or meddled with. In the history of neuroscience, patients who suffered brain injuries or underwent brain surgery have provided insightful, fascinating stories that illustrate how lesions in specific cerebral regions may result in marked alterations in behaviour. Some are particularly telling and memorable.

By far the most famous and most often recounted such story concerns Phineas Gage, a 25-year-old American who in the mid nineteenth century worked as a railway construction foreman and suffered an unfortunate and unusual accident in the course of his duties. Since new rail lines needed to be laid across the state of Vermont, it was essential to flatten the uneven ground and Gage was responsible for carrying out controlled explosions. The procedure was relatively straightforward: Gage had to first drill holes in the ground, fill them with dynamite, insert a fuse and lastly push a tamping iron down the holes after the explosive powder had been covered with sand. On 13 September 1848, because someone called him and he briefly turned round, something in the protocol went wrong. Gage started to tamp before one of his assistants had applied the sand. This was a grave mistake, because, without the sand, the explosion spreads away from the rock. The result was that the tamping iron, over a metre long and three centimetres thick, blew out and went right through his head, exiting from his left cheek, before it rocketed into the sky and fell to the ground several yards away, leaving everyone present astounded.[11]

It's hard to believe, but Gage survived. Remarkably, after momentarily losing consciousness, he regained it immediately after the accident. And, after a few weeks of convalescence, he recovered fully. His language and intellectual capabilities were entirely unaffected. He could walk, run, talk and interact with people and even go back to work. Over time, however, everyone noticed a few changes in his personality.

Before the tamping iron penetrated his skull and brain, he was unanimously regarded as a considerate, loyal and friendly man by his peers. At work he was praised as one of the best and most efficient workers, the company's favourite. However, after the accident and as early as his convalescence, he had bursts of anger, became impertinent and impulsive and lost his capacity to judge the social acceptability of certain of his ways of behaving. He became unreliable, offensive and irresponsible towards others.[12] Eventually Gage was left isolated by his friends and acquaintances. He lost his job and never found another. Having descended into a desolate existence, he died a dozen years later.

This tragic story is scientifically compelling in that it demonstrates the links between brain damage and behaviour, in particular social and moral behaviour.[13] Gage's case showed that compromising a fraction of the brain can have serious and noticeable consequences on a man's personality. His skull and the infamous tamping iron remained on display at the Warren Anatomical Museum at Harvard University and, remarkably, for a long time they did not receive the attention they deserved. In the mid 1990s Antonio Damasio and his colleagues at the University of Iowa College of Medicine decided to examine the skull to reconstruct the accident and closely map the brain areas where the lesion occurred. They established that the tamping iron had specifically damaged the ventromedial part of the prefrontal cortex. This was an important clue. Damasio had met other patients with similar lesions and comparable behaviour. So he set about investigating them.

One of the group's first experiments that helped identify the role of emotion in decision-making focused on gambling. Not everyone is a professional gambler, but we all face decisions that require the assessment of risk and of potential gains and losses, as well as choices that may conceal harmful, counterproductive and irreversible consequences. Such are the uncertainties of life.

Damasio and his colleagues gave the players in the gambling

experiment a starting sum of $2,000 and four decks of cards, asking them to draw from any of the four decks.[14] Each card drawn revealed a reward or a request to hand over an amount of money. The ultimate goal was to end the game with the highest profit. A secret pattern lurked among the cards. One pair of decks contained cards with the best rewards, up to $100. However, this pair also included cards that requested the gamblers to hand over equally large amounts of money. So, while these two decks gave the impression of being profitable, they also carried the highest risks. At first sight, the gamblers had no way of telling when an unfavourable card would turn up. By contrast, with the other two, less treacherous, decks of cards, the highest win was only $50, but the losses were never harshly punitive. Overall, drawing from the low-win decks would prove more profitable.

The gamblers in the experiment consisted of two groups: people with their brains intact and patients with lesions in their medial prefrontal cortex. Like Phineas Gage, the latter experience difficulties in taking decisions. Damasio realized this when, for instance, he invited them out for lunch and asked them to pick the restaurant. Testing Damasio's patience, they would spend more than half an hour reciting the pros and cons of several restaurants. One, they warned, had good prices but was always empty, so it might not be too good, but, on the other hand, it was more likely to have a free table; another was pricey but had generous portions.[15] In the end, despite all their lucubration, the patients couldn't make up their minds. One of them, whom Damasio named Elliot, was a bit like Gage. He was an otherwise entirely intelligent, pleasant and charming man with a sharp memory, but he was unable to hold on to a job, keep a wife or plan his time properly. He acted foolishly and irresponsibly and could not be trusted.

Anyway, back to the experiment. As the gamblers carried on playing, an important hint that made Damasio suspect the involvement of some kind of emotional arousal in their choices came from their bodies – to be precise, from their skin. Attached to each gambler's skin was a machine that measured changes in skin conductance response, or SCR. SCR is a sophisticated expression for sweat. If you are nervous or stressed, or in general emotionally stimulated, one of the things that happens to your body, even if it is not perceptible to the naked eye, is that your skin sweats slightly. In a laboratory, this can be measured as it happens. Over the course of the game, the gamblers with intact brains preferred to pick cards from the advantageous decks. At a conscious level, they didn't know exactly what was going on or why it would be wiser to take that decision. But their bodies did. As measured by the SCR, each time they picked from the risky decks, fear emanated from their skin and that emotional edge guided their choice towards the less hazardous decks. On the contrary, as you would expect, the judgement of the patients was less sharp. When their hands reached for the more punitive decks, there was little or no skin reaction. They kept drawing cards from the bad decks, even when they started to realize how harmful they could be.

So, failure to experience the emotional cues of a situation results in poor deliberation.

Not only was emotion important in guiding a decision, but in a way it already knew which was the best decision to take, and took it first. Call it intuition, a sixth sense or just plain foreboding. Whatever it is, it helps reason to make a choice.

Damasio's hypothesis is that this intuition is actually finely etched in our brains, like grooves of a song incised on a vinyl record. In fact, he calls it the 'somatic marker hypothesis' (the Greek word *soma* means body). Each time we face a situation, we register its positive or negative emotional charge. It's as if we stored emotional knowledge in our brain. The behaviour in the game of the two kinds of gamblers suggested that the acquisition of this knowledge must somehow require a functional prefrontal cortex – in connection with the limbic brain – and that, in possession of this knowledge, the prefrontal cortex works like a guide that controls our actions. Indeed, the acquired information becomes precious knowledge for when a similar situation arises again. The harsh losses they incurred taught the gamblers with intact brains about the risk of drawing from the bad decks. The gamblers with lesions in their medial prefrontal cortex could not register, nor retrieve, that information, and so kept making the same mistake.

In real life we face countless situations in which emotional knowledge comes in handy. These range from relatively simple choices, such as which colour to paint the living room, where to spend a holiday or which painting to buy, to more committed decisions about who to date, which property to buy or whether or not to accept a job offer. In each of these cases, emotional hints can guide our actions. It's almost as if the grooves of that once-incised song play a warning sign silently in our ears, suggesting what we should do.

Damasio's ground-breaking experiments entirely revised the predominant theories that confined decision-making to the realm of rationality and established a new theory according to which emotion is essential in decision-making and our most seemingly rational choices. Emotion and reason are not two exclusive functions of the brain. There exists a mutual dependency between the two. Relying on the computational qualities of your brain makes you develop sophisticated analyses. But, as Damasio's experiments show, you would not be able to take any good decision. In extreme cases, no decision at all. You would be blocked or lost in the careful assessment of the myriad advantages and disadvantages of each option, just like those patients who couldn't make up their minds about the restaurants. It does happen from time to time that we take decisions without being able to provide the ultimate explanation for having taken them. Emotion helped us take them, unconsciously, behind the foreground of rationality. So, emotion makes its own judgement, as it were, and has equal authority to rationality. In fact, reason can't operate without emotion's persuasive advice.

But what these experiments also did was to remap the fixed geography of brain function. They showed that a region in the prefrontal cortex, which everyone believed was exclusively responsible for the analytic, logical duties of the brain, does indeed participate in emotion. Without it, the emotional edge that contributes to decision-making somehow can't be integrated into the process.

After Elliot, several other patients were observed in a search for clues that could confirm the original findings.[16] In some cases the lesions in the prefrontal cortex resulted in syndromes characterized by pronounced aggression and impulsivity. A 56-year-old man with the initials J and S – I'll call him Jay – was taken to a London hospital's emergency department after he was found unconscious with damage to the front of his head.[17] An inspection of his brain revealed damage to the orbitofrontal cortex (as well as to the left amygdala), the lowest and most frontal part of the PFC, behind the eyes. His behaviour became bizarre while he was still in hospital: for instance, he was found riding a hospital trolley. Like Gage and Elliot, he failed to plan ahead properly, sometimes taking trips around London without any particular destination or any idea of when he would come back. He was also unable to hold down a job. Basically, damage to Jay's PFC compromised his ability to plan, keep things in mind and pay attention. But he was also irritable and aggressive. He became uncooperative with the hospital staff, whom he frequently assaulted and wounded. He had lost all sense of what could be dangerous for others. He showed no respect for the safety of those around him and no remorse or guilt for his actions, even when he hit nurses. On one occasion he kept pushing a patient around in her wheelchair, despite her screams of protest. He wasn't sensitive to clues about the social acceptability of his behaviour, nor did he accept responsibility for his actions.[18]

Further evidence of the role of the PFC in controlling aggression was found in a group of murderers who had committed unplanned, impulsive murders. Their brains showed abnormalities and decreased functionality in various areas of the PFC.[19]

Evidence accumulated from observation of several individual patients points to the PFC normally performing an inhibitory function on tissues of the limbic system, including the amygdala.

However, when the PFC is lesioned or something else goes wrong with it, the amygdala is released from this inhibition, making it harder to control aggression.[20]

The general picture that comes out of this research is that the prefrontal structures exert a regulatory or modulating role on the limbic regions. The prefrontal cortex constrains impulsive outbursts. This is possible because these two systems are not isolated from each other. On the contrary, they are delicately connected to allow the integration of their functions. The ultimate outcome of an action must be finely tuned by both the limbic structures and the prefrontal structures.

The establishment of control, and the wise use of restraint, may come in handy across a spectrum of actions that call for it, from the most trivial choice to the most despicable act of violence. For instance, it is thanks to the prefrontal cortex that we resist the temptation to spend money we don't have, or opt for a sugar-free coffee to minimize glucose intake with the aim of preserving our figure.[21] Without the PFC, we would have a hard time completing a task. We would also be indifferent to the good or bad value of things. Or we couldn't restrain our anger.

## An angry bunch

There is another level at which people differ in the way they develop and manifest anger and violence. From the anatomy of the brain we need to move down to something invisible: genes.

Genetics is all about looking for differences. To learn about the function of a gene, geneticists study what happens when something goes wrong with it, when it is absent or when it has undergone changes, or, in biological parlance, mutations. A strong clue to a genetic component of aggression came from the Netherlands. A group of men from the same large family presented persistent and pronounced aggressive behaviour.[22] They displayed an elevated predisposition to aggressive outbursts, excessive anger and violent, impulsive behaviour, such as rape, assault and attempted murder, burglary, arson and exhibitionism.[23] A few also presented mild mental retardation. The fact that the trait kept manifesting in the same family made Hans Brunner, a scientist working in Amsterdam, suspect that their behaviour might have been the outcome of some anomaly in their genetic make-up, so he set out to sequence the men's DNA. What he found was remarkable. All of them carried a faulty version of a gene responsible for the production of an enzyme called monoamine oxidase A (MAOA). The mutation was in their X chromosome, the genetic material we inherit from our mothers.

Among other things, enzymes break down other molecules. MAOA breaks down neurotransmitters, such as dopamine, norepinephrine and serotonin – molecules that allow brain cells to communicate with each other – all contributing in one way or another to the quality of our moods and personality. The Dutch men's mutation was an infrequent but rather powerful anomaly. Basically, these men did not produce any MAOA.[24] After this rare discovery, more scientists looked into whether other versions of the MAOA gene existed in the human population.[25] While the sequence of genes across individuals is pretty much identical, there may be tiny differences at the level of the DNA bases – the units that make up a DNA molecule – that make each of us unique and different from everyone else. These differences constitute what is called genetic variation. Often these changes are without effect. Sometimes, however, they result in the alteration or loss of the functionality of a molecule.

Indeed, in the population at large there is genetic variation for MAOA; that is, there are slight differences from one individual to another in the relevant DNA sequence of that gene. The MAOA gene comes mainly in two forms: a longer version producing high levels of the enzyme and a shorter version producing low levels. If you have less enzyme, there will also be less effective and slower degradation of neurotransmitters in your brain. In one study conducted in 1993, men with the low-activity version were found to be more likely to engage in impulsive and aggressive behaviour. As additional evidence, rodents whose MAOA gene has been engineered out have elevated levels of serotonin and males manifest a dramatic increase in aggressive behaviour.[26]

After the discovery of its implication in aggression and violence, the MAOA gene was rapidly given the nickname the 'warrior gene' and a flurry of articles have been published all claiming association of the low-MAOA form with aggression and violent behaviour, as if aggression and violence could be the result of bad genes only.

In the 1990s, when these discoveries were made, there was great excitement about the role of genes and their influence on behaviour. Well over forty years after the discovery of the structure of DNA in 1953 and the realization that this molecule was the carrier of the genetic information, the global scientific community was working towards the next big milestone: decoding the genome, that is the sequence of an individual's entire genetic material. With the race on to complete the Human Genome Project, you could breathe the enthusiasm in laboratories. Genes ruled.

Press reports full of bad popular science contributed to the spread of the simplistic notion that for every behaviour there was a gene, and that it could be discovered. This kind of talk was labelled 'genetic determinism':[27] the belief that we are destined to behave in certain ways because of our genetic make-up and neuronal wiring. However, very soon after the publication of the Human Genome, it became clear that, for complex behaviour, the effect of genes was relatively small. You are not violent because you carry a particular form of a gene. A direct causal relationship between genes and behaviour is valid only in some instances, when a single gene-defect leads to brain dysfunctions.[28] A classic example of this kind is Huntington's disease, a neurodegenerative disorder that causes nerve cells to waste away, resulting in poor muscle coordination and dementia. If you happen to have, in chromosome 4 of your genome, an excessive repetition of a short DNA sequence, called a CAG repeat, no matter what you do, where you grew up or where you live, you will develop Huntington's.

However, the origin of most behavioural traits is more complex than that. For one thing, most traits are 'polygenic', in that they involve the concerted interplay of many genes at the same time. MAOA is, so far, certainly the most studied and most credited gene with a link to aggression, but it is not the only one. What makes matters more complicated is that one gene can be responsible for more than one behaviour. So, while we refer to the 'gene for' Huntington's disease, it is not correct to allude to the 'gene for' a complex trait, such as aggression. In fact, MAOA could be given even more labels. It could be called the 'depression gene' or the 'gambling gene', because variation at the level of its sequence has been found to be present in individuals manifesting those behaviours.[29]

Taken alone, knowing which variation of a gene a person carries is useless in predicting whether he or she will manifest a particular behaviour. Many more variables are involved.

## Genes and environment

One of these variables is unquestionably the environment. Behaviour can't be studied without an appreciation of the circumstances in the external world where it manifests and which contributes to its emergence. Upbringing and traumatic experiences have strong effects on the development of personality. The environment interferes with the action of some of your genes and compromises the outcome of your development. For instance, identical twins who have exactly the same genome may end up with dissimilar personalities if reared in different families or communities.

In the case of antisocial and violent behaviour, factors as diverse as childhood abuse or neglect, unstable family relationships or exposure to violence have all been found to be influential. A good proof of that came from a ground-breaking study conducted in New Zealand by a team led by Avshalom Caspi and Terrie Moffitt. Together with their colleagues, they set out to investigate whether variation in the MAOA gene could modulate the effect of these various kinds of childhood maltreatment. The researchers were lucky to have access to a cohort of people whose lives were progressively monitored from the age of three to twenty-six through surveys, family reports, tests and interviews. As best they could, they basically kept track of how the study participants grew up and led their lives. They found that, although MAOA alone had no large effect, it definitely modulated the impact of early-life maltreatment on the onset of antisocial behaviour, with people carrying the low-activity form of the gene being significantly more susceptible to the effects of abuse than those with the high-activity form (Fig. 3).[30]

Over 80 per cent of those carrying the low-activity form ended up developing antisocial behaviour, but only if they had been exposed to maltreatment and abuse during their lives. By contrast, only 20 per cent of those carrying the malfunctioning form of the enzyme became violent if they had grown up in a healthy environment, without maltreatment.

Subsequent studies have independently come close to the same

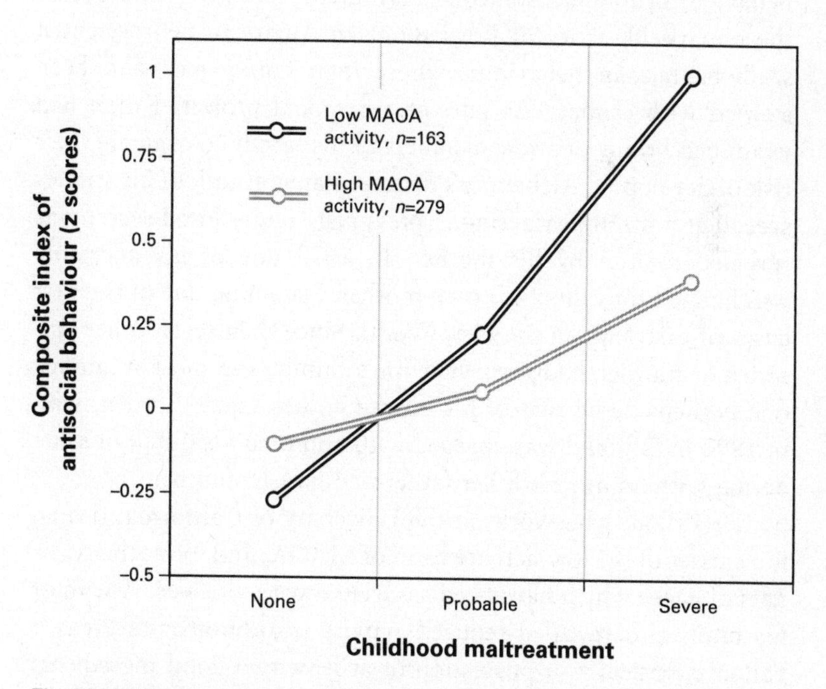


Fig. 3 MAOA gene–environment interaction. After exposure to severe forms of maltreatment in childhood, individuals carrying the low-activity form of MAOA are more likely to manifest antisocial behaviour. (From Caspi *et al.*, 2002, reprinted with permission from American Association for the Advancement of Science)

finding and tested other forms of environmental influences and measures of violent behaviour, including self-reports of aggression.[31]

The overall message to bear in mind is that a gene alone does not translate into an emotion. A gene is not the *essence* of a behaviour. MAOA is not a synonym for either aggressive behaviour or criminality. The reason why genes are important and scientists keep hunting them down is that identifying a gene offers enticing clues about the general mechanism of a behaviour, particularly one that has clinical consequences. By finding a gene you can locate the neurochemical pathway that contributes to the manifestation of the symptoms and, of course, where in the brain the behaviour or disease is likely to be mapped.

However, no neuroscientist would ever tell you that variation in a gene such as MAOA is alone sufficient to determine violent behaviour or to make someone a criminal. Recently I came across the remarkable story of Jim Fallon, an American neuroscientist studying human behaviour, whose own family past had been stained with crime.[32] As part of a personal project, Fallon had examined brains of a few members of his family to evaluate their risk of developing Alzheimer's disease. Later, as talk of his studies spread at a family gathering, a previously undisclosed secret was revealed to him by his mother. In 1673 one of his ancestors was hanged for killing his own mother, becoming one of the first cases of matricide in the New World. Since then seven other episodes of murder had tarnished Jim's family, the most infamous one perhaps being that of his distant cousin Lizzie Borden, who in 1892 in England was charged with and then acquitted of murdering with an axe both her father and her stepmother.

Dr Fallon, who works at the University of California, Irvine, is a carrier of the low-activity form of MAOA, and four other variants of genes which have been associated with violence. A scan of his brain also revealed reduced activity in orbitofrontal areas.[33] Fallon is basically in possession of at least two good ingredients that could potentially make him a violent killer. The chances that he would become one were greater than for other people lacking those biological attributes. Yet he did not. Apart from a tendency to indulge in risk-taking behaviour – like trout fishing in a corner of Kenya frequented by lions – nothing in his behaviour points to a threatening, violent attitude. Why? As Fallon himself explained, one essential ingredient was missing in the recipe for violence: a bad childhood. He says that as he grew up he did not experience trauma, nor was he exposed to any hostile environment. His was an easy childhood. Obviously, a deeper and more detailed investigation of his brain and life history, as well as those of his relatives, is warranted, but this interesting episode in the life of a neuroscientist studying behaviour shows the relative power of genes.

During the day I spent in the countryside I got to know Bruce a little better. He insisted that his DNA be tested. However, I was then able to persuade him to do more justice to science by enrolling in a study that involved a few hundred participants and measured the extent of their aggressive behaviour in relation not only to DNA variation, but also to information about childhood, upbringing and life history – all kept anonymous.

I tested my own DNA and, please permit the disclosure, I carry the high-activity form of the MAOA gene, so it seems consistent with the fact that I am not particularly prone to develop or manifest anger – to some extent, of course, because of the healthy environment where I grew up. But even if I were carrying the low-activity form it wouldn't necessarily make me violent, any more than it did Fallon. It is the presence of the gene in combination with a hostile environment that increases the possibility of developing antisocial behaviour.

## The brain on the stand

Ever since the discovery of a link between genes and aggression, lawyers have attempted to use such biological information as evidence that might justify their clients' criminal actions on the basis that their *bad* genes or brains made them commit the crime.

Though never immune to imperfections, the system of justice follows a rather straightforward course. A suspect is charged with a violent crime. If they are found to have committed the crime, and voluntarily so – that is, with a guilty mind – they will be sentenced. An offender who is not in full possession of his or her mental capacities is accorded a lighter judgement. The task of ruling with certainty on a suspect's mental capacity is a significant challenge for judges and medical experts alike, and the practice and outcome of such deliberation have depended on the available medical knowledge at any given time in history.

Until not long ago culpability in suspects with possible mental problems was ascertained solely on the basis of extensive psychiatric evaluations. Today the introduction of genetics and neuroscience into the courtroom shakes established notions of agency and culpability.

The first case in the world in which MAOA was used by the defence as a mitigating factor dates back to a 1994 US trial. Since then genetic evidence has been used worldwide in at least two hundred cases, of which about twenty were in the UK.[34] In 2009 a court in Italy cut the sentence to a convicted murderer by a year because he carried the low-activity version of the MAOA gene.[35] This became the first case in Europe in which genetic information affected a judicial sentence. The murderer was Abdelmalek Bayout, an Algerian citizen who stabbed and killed a man who insulted him about the kohl eye make-up he was wearing for religious reasons. In his verdict the judge who mitigated the sentence stated that he had found the MAOA evidence particularly compelling and embraced the motive put forward by the forensic experts, who claimed that Bayout's genes would make him behave violently if provoked. In the US, even brain imaging has been introduced to alleviate the culpability of a defendant, but this has not been used in UK courts.[36]

In early 2012 an interesting and informative survey of almost two hundred trial court judges in the US revealed that expert testimony providing biological evidence led judges to impose more lenient sentences when asked to deliberate on an offender in a fictional case of battery that was inspired by a real event.[37] On average, the judges cut the sentence by one year. However, the respondents to the survey disagreed on the weight that should be given to the biological information – which included MAOA genetic evidence as well as atypical amygdala function. For some, the biological evidence was a mitigating factor, because it represented an immutable, intrinsic cause for a behaviour over which the offender had no control. Interestingly, another group of judges argued the opposite and held the view that offenders with risky genes and risky brains would be a constant danger for society, declaring them prone to reoffend and unable to learn from punishment. This latter group of judges was more concerned with the future than with the past actions of the offenders. They didn't feel comfortable with giving them back to society sooner than necessary.

The neuroscientist and author David Eagleman has been a hopeful proponent of the possibility of using neuroscience in the courtroom. He argues that the current legal notions of culpability and blameworthiness are bound to evolve in light of progress in neuroscience.[38] Whether it is a change in brain morphology, a clear genetic defect or a more subtle neurochemical alteration, there will always be a biological explanation for a criminal's bad behaviour and such explanation will have to be taken into account during deliberation on a sentence. As a result, notions of volition, free will and blameworthiness will undergo transformation. For Eagleman, the question of blameworthiness is the wrong one to ask in the legal system, because in time neuroscience will reveal what elements in the brain biology of every offender can make him or her perpetrate a crime. A sentence given today to someone deemed culpable of committing a crime may change in a few years because of new ways to assess the biology of his or her brain. In line with the forward-looking judges in the survey, Eagleman concludes that the right question to ask is, how likely are the criminals to offend again, on the basis of their biology, which we will progressively understand better.

On 19 July 2012, James Holmes, a 24-year-old student who had dropped out of a neuroscience doctoral programme, opened fire in the darkness of a cinema in Aurora, Colorado. His target was an innocent audience attending the premiere screening of *The Dark Knight Rises*, the third film of the Batman saga. Holmes carried a Remington 870 shotgun and an assault rifle, and wore an oxygen mask and a Kevlar suit that made him look like the movie's evil villain Bane. When Holmes threw a smoke bomb, some of the witnesses who survived the rampage said that at first they thought it was all part of the spectacle of the premiere, believing the man in disguise was an enthusiastic Batman fan dressing up like one of the characters in the film.[39] Holmes killed twelve innocent cinema-goers and wounded fifty-eight. He was caught and still awaits judgement. At the time of the crime, Holmes was in therapy with a psychiatrist, and he tried to reach her on the phone just minutes before starting his rampage.[40]

Unfortunately the Aurora shooting was not an isolated event. In the US, in 2012 alone, several similar events preceded and followed Holmes's attack. In June 2012 a gunman shot three people at a pool party near the campus at Alabama's Auburn University. Two weeks after the cinema rampage in Aurora, a man killed seven and wounded three at a Sikh temple in Oak Creek, Wisconsin. In December 2012, just eleven days before Christmas, 20-year-old Adam Lanza carried out one of the most atrocious and deadly rampages ever witnessed on a US school campus. He opened fire on innocent staff and children in an elementary school in Newtown, Connecticut, after killing his own mother at home, murdering twenty-eight people in total.[41] Twenty of these were children between six and ten years old. The toll of victims at Newtown's elementary school is second only to the shocking loss of lives caused by the shooting at Virginia Tech in 2007, which left thirty-two people dead. And, of course, everyone will remember the 1999 massacre at the Columbine High School in Colorado.

While neuroscience refines its tools to understand the biological basis of violence, it will always be useful to keep an eye on how society deals with crime, and with mental pathology. Ever since the first links between genes and behaviour such as aggression were discovered, a few intellectuals – including scientists – have voiced their concern about the danger that granting genes and brains the exclusive power of governing behaviour would exempt us from critically assessing and modifying some of society's policies that may contribute to aggressive and violent behaviour. For instance, if we really believed that genes are all it takes to mould intelligence, there would be no reason to invest in improving our systems of education or in promoting culture. Similarly, the identification of the biological components of aggression and violence has somehow shifted the attention from some of the social factors that contribute to their rise. An equal worrying consequence is the tendency to misunderstand mental illness in general.

In the weeks following the Newtown shootings, geneticists set out to examine Adam Lanza's DNA to screen for the presence of anomalies in its sequence or for any variation that could be linked to violence.[42] As yet, no results have been revealed. However, it is unclear how the information obtained would be used and for what purposes. One guess is that if any conclusive information is gathered, it might be used to screen the population for the same anomalies and prevent future crime by identifying potential offenders in advance, even among school children.[43] But this is not a straightforward undertaking. There is no doubt that genetic variation shapes our brains and that our neurotransmitter levels fluctuate during aggressive reactions. Yet granting such genetic changes the power to directly cause particular behaviours or decisions requires careful consideration. In the case of MAOA, for instance, it would mean that all those carrying the low-activity version of the gene should be given a shorter sentence for their crimes, but it is certainly not the case that all people carrying the low-activity version go around attacking people. To gain a more precise perspective, it is useful to bear in mind that the prevalence of the low-activity form of the MAOA gene, at least in Caucasian populations, is 34 per cent. This means that in such a group about one in three individuals carries the low-activity form, but certainly not one in three goes around committing crimes.

Launching prevention campaigns among the population would surely create stigma. As we have seen, the environment alone plays an enormous role in the increase of violence. An upbringing marked by hostility, and factors in a person's life trajectory such as abuse, abandonment and, in general, a violent environment are often a prelude to the onset of violence. Genes are only modulators that can either magnify or attenuate the effect of these, like the volume knob of a hi-fi. There is something else that can be done in parallel or instead of screening for DNA mutations. That is to invest in successful social welfare programmes.

We may peer into the brains of violent perpetrators to look for anomalies in the prefrontal cortex. We may even check their genotype for MAOA and various other genes. But every brain is different and every brain changes constantly. So, in order to find the exact physiological conditions that made someone commit a violent crime, we would have to inspect their brain at the time of the act.[44]

Finally, let us not forget that, at least in America, individuals like James Holmes and Adam Lanza, as well as all those with a malfunctioning prefrontal cortex or the low-activity version of the MAOA gene, could not commit crimes if there were stricter regulations for purchasing guns or rifles.[45]

Only days after the Aurora killings, terror spread on a crowded Manhattan sidewalk on West 33rd Street and Fifth Avenue, close to the Empire State Building. A man pulled a gun to shoot his former employer who had sacked him one week before. As reported in a *New Yorker* article following the Manhattan shooting, at the end of a press conference held to brief the public on the events, New York City mayor Michael Bloomberg said laconically: 'There's an awful lot of guns out there.'[46]

## Pacify your frustrations

I have talked at length about anger as a prelude to unacceptable, deplorable violence, as a negative emotion to be avoided and kept at bay. But anger is not always followed by aggression. Violence can also erupt in the absence of anger. Psychologists and philosophers debate the benefits of ignoring anger in the attempt to stay cool, rather than venting it. As Aristotle said in his *Nicomachean Ethics*, anybody can get angry. But expressing anger in the right tone, at the right time and for the right purpose demands careful judgement and a touch of virtue. This is an ability that we begin practising as kids, when we need to learn how to react to the first forms of injustice – when, say, somebody bullies us, or a school mate nicks our brand-new pencil – and then gets polished over the years, when we become adults and, hopefully, reach some level of wisdom, though I believe we never stop learning.

Sometimes, banging your fist on the table or clearly remonstrating is better than allowing resentment to brew inside, and may prevent you from ending up taking unpleasant actions.

Both spontaneous outbursts of anger and anger that stews inside us can have serious repercussions on our health. Primarily, anger has a toll on the mechanics of the heart. There have been studies clearly showing that reacting to stressful situations with anger increases the risk of premature cardiovascular diseases, particularly myocardial infarction.[47] On the other hand, letting out anger constructively, especially in everyday episodes that don't escalate into aggression, has positive consequences.[48] If our anger is justified, lucidly expressing the reasons for it can improve relationships and lead to healthy solutions that benefit all parties involved. So, it's worth striving to stay within a moderate threshold of anger.

Knowledge of the brain circuits governing emotional control has spawned the development of techniques that aspire to teach us how to quench or control our anger by looking inside our brain. In the near future, such self-control may be gained by this informed taming of the brain. David Eagleman calls it the 'prefrontal workout' and, as you would expect, it has to do with exercising the regulatory power of the frontal lobes.[49] The technique would consist in watching on a screen the activity of your brain circuits when you are fighting the temptation to indulge in something you know is bad for you, like eating chocolate cake, or when you are trying to avoid bursting out in anger. As you restrain yourself, you watch a bar that signals the involvement of your frontal circuits and the achievement of control. If it stays high, you need to work harder. As you concentrate to tame your urge, you learn which mind strategies help you bring the bar down, and the corresponding brain circuitry will be trained to achieve the desired goal. If such techniques find concrete application in the years to come, it is imaginable that they could be applied to the rehabilitation of offenders, as a parallel or even alternative solution to imprisonment. This sounds like a much less disturbing version of the Ludovico Technique, the therapy used on Alexander DeLarge, the protagonist of *A Clockwork Orange*, that conditioned him to feel nauseous each time he witnessed or even thought about committing violence. With the help of a pill, DeLarge was taught to feel sick while watching scenes of violence. In the 'prefrontal workout', one would actually teach the brain to abstain from violent behaviour.

Almost two millennia ago, the ancient Roman philosopher Seneca wrote an entire book on anger and came up with a smart approach to avoid it. Seneca knew well that anger was an inevitable component of existence. He lived all his life in ancient Rome, which, even then, was not the calmest place on the planet. 'If one runs off on many different activities, one will never have the luck to spend a day without some annoyance arising, from someone or something, to dispose the mind to anger.'[50] If we venture into crowded areas in a city, it is likely that we will bump into many people, or that someone will step on our feet. In life, something will always go the opposite way to what we would like. Plans do not always take the course we anticipate: 'No one has fortune so much on his side as always to answer to his wishes . . . ' said Seneca. Indeed, it is extremely easy to lose one's temper and become angry at the person or the situation that provoked the annoyance, and even at oneself and one's bad luck. Yet, for Seneca, anger was demeaning and was best avoided. 'It is not how the wrong is done that matters, but how it is taken.' For Seneca, it was important to take one's time to examine the real nature of the annoying incident or situation and, above all, to avoid falling prey to provocation: 'Beyond any doubt, one raises oneself from the common lot to a higher level by looking down upon those who provoke.'

## Coda

Much of what we know about the biology of disinhibited behaviour, aggression and violence has emerged from the singular stories of individuals whose observed actions, localized brain lesions, genetic deficits and life vicissitudes have contributed to the sketching out of a preliminary physical map of emotion regulation. From the almost legendary Phineas Gage and Damasio's improbable patients to Jay's bizarre behaviour and the criminal actions of Abdelmalek Bayout, James Holmes and Adam Lanza – and even Bruce's impatient impulsive reactions in the car – we have seen anger and the loss of emotional control in various complexions. Like obscure characters in crime fiction, these figures traced their own life paths. Each of them was or is a unique individual with distinct intentions, motivations and values. They each possess a brain that bears the signature of their own past. They display behavioural similarities and differences and share biological features, but they all retain a degree of individuality. Gage's brain is slightly different from Elliot's, which in turn differs from Jay's. Gage and Elliot did not become criminals. Jim Fallon and Abdelmalek Bayout both carry the low-activity version of MAOA, but Fallon never committed a violent crime.

The stories of the characters in this chapter have shown how specific abnormalities in the brain and the genome have tangible, sometimes dramatic, effects on behaviour. Yet the overall essence of each individual and what makes them who they are is the outcome of a vast and complex set of contributing factors – all in conversation with their biology – that we are only beginning to understand.

Our brains, and more generally our whole bodies, are the physical substrates of our actions. However, they don't simply work in complete isolation from the intricate interpersonal, social and historical contexts in which we live.

The neuroscientist Steven Rose offers a fascinating vision of human beings as living organisms who build their life trajectories through time and space and according to their biology. He recognizes the power of genes and our material selves without subscribing to determinism. We are not slaves to our genes. Rose calls such trajectories 'lifelines', for they are like paths we construct and decide to follow.[51] As we move along these trajectories we may in time narrow the distance between the behaviour we display, the choices we make, the feelings we have, and what we know about what goes on in our brains. The concluding message of this chapter is that behavioural features emerge from a biological architecture that makes them possible, and whose variation gives individuals personal and unique shadings of those features. The truth is, however, that our every action can be explained at multiple levels, from individual neuronal firing and chapters in our biographies, to environmental circumstances and social contexts.

# 2

# Guilt: An Indelible Stain

Guilt has very quick ears to an accusation

HENRY FIELDING

A good deed never goes unpunished

GORE VIDAL

The window was semi-open. The early-morning sun rays flickered through the slits in the venetian blinds that kept banging against the window, at the mercy of a mild, yet insistent wind. For a few moments, I wasn't sure whether I was awake or still sleeping, lying in the borderland between fact and reverie. I remained motionless, trying to make sense of my surroundings. I had forgotten where I was. An unpleasant feeling had stirred me in the early hours of the morning with a dream, which I was determined to remember, without letting it dissipate through the sieve of my consciousness. It was a rather curious dream. At the centre of it was a date I had made with my old friend Esra to see each other in Rome, something we had looked forward to for a long time. On the day of our appointment we arranged to meet on the river bank, close to her hotel in Trastevere. When I arrived at our agreed meeting point she wasn't there and I sat by myself on a bench. While I waited, a few figures stopped, one after the other, to ask me for the time and what day it was. A beggar, a traffic warden, a policeman, even a nun. Each time, I looked at my watch and I obliged with an answer, after which they all ran away fretfully saying how late they were. No sign of Esra, though. Time goes by erratically in a dream, but the wait felt like an eternity and started to make me impatient. I called the hotel, but no one answered. I tried her mobile and she would not pick up. I slowly became bored and a little upset. Then a line of professors holding glasses and microphones paraded in front of me. They all stared at me and I didn't understand why. Some were inquisitive, others impassive. I left a message on Esra's voicemail. All of a sudden, I began to hear a loud, banging sound as if something kept being dropped from the sky to hit the ground. I worried something might have happened to Esra, but I also lamented her standing me up. I called again, in vain, and left another message. Finally, I rose from the bench, trying to locate the source of that sound. I turned around several times, but there was nothing to see. Then I woke up, that annoying sound echoing with the irregular bangs of the blinds against the window.

You may be wondering what that dream was all about. I more or less knew what it signified the minute my eyes opened. Those seemingly absurd figures, and the bizarreness of their actions, the long wait and the intrusion of associations to time and the disappearance of my friend, were the disguise for something that was troubling me: guilt.

For several weeks, I had carried with me an unpleasant sensation that I had more or less successfully put aside thanks to the conveniently distracting thrust of daily routine, which is commonly deft at burying emotions. On a brief holiday, that sensation found the path to re-emerge. Someone was knocking on the door of my conscience. The truth was that a few months before, Esra had invited me to speak at an interesting conference she had organized. Flattered by the invitation and excited by the opportunity, I enthusiastically accepted. But I sloppily failed to mark the date in my calendar! Busy and overworked, I completely forgot about the invitation. Then, just a couple of weeks before the conference, came a gentle reminder to confirm my participation and submit my paper. What?

Panic.

I was supposed to give a paper I had never given before, and a few other trips and speaking commitments stood ahead of Esra's symposium. Even if I had decided to do without sleep from here on to the deadline, it would have been impossible for me to be ready for the occasion and honour the invitation with a decent lecture. Reluctantly, but with no better choice, I cancelled, with endless apologies. But Esra wasn't pleased with me at all. Understandably. I was afflicted with guilt. I felt awfully bad that I had not been able to fulfil my commitment and I couldn't believe I had neglected to meet the request for my participation in the symposium, especially since it came from a friend. I have organized conferences myself and I know what it means to find oneself with an empty slot at the last minute. I was haunted by negative judgements about my conduct and deeply hated myself for not doing what I should have done – marking my calendar, keeping track of my schedule, preparing and honouring my friend's kind invitation.

Like a ghost, guilt often materializes in dreams, disguised in more or less inscrutable, at times bizarre permutations. It was, in fact, a guilt-themed dream Sigmund Freud experienced himself in the summer of 1895 that helped him formulate his theory on the interpretation of this enigmatic nocturnal stream of unconsciousness.[1] In Freud's dream, everything pointed to a sense of guilt he felt for a misdiagnosis of a patient, Irma, who was also a friend of his. According to Freud, Irma suffered from hysteria. After a period of treatment, Irma got better but she kept experiencing somatic pains and unease. Freud, however, discounted her medical symptoms and established that what she was experiencing did not have an organic nature.[2] On the evening before Freud had the dream, Otto, one of his best friends, who had recently visited Irma, reported she was better, 'but not altogether well'. Freud sensed some kind of criticism hidden in the tone of Otto's voice and interpreted his remark as a reproach, perhaps a message coming from Irma and her family, for the superficial therapeutic choice he had made. Freud became upset about this. The dream has for its setting a party at his house, at which Irma is also present. In the initial moments, he takes Irma aside and tells her bluntly: 'If you still have pains, it is really only your own fault.' Freud then examines her throat, which he finds to be full of greyish and white scabs, clearly proving the presence of an infection, which was also confirmed by another doctor present in the dream. Irma had in reality received an injection and in the dream Freud suspects that perhaps the injection had been carried out sloppily and with a non-sterile syringe.

Freud clearly felt responsible for having underestimated Irma's condition, but he shifts his own blame on to her and the other doctor for mistreating her. The experience is so strong and the guilt so unacceptable that he sheds it on to others. But he knew all too well that in fact the dream was about his own discomfort with the failure, real or perceived, of his treatment of Irma. Thanks to this revealing experience, he concluded that 'the dream has a meaning, albeit a hidden one; that it is intended as a substitute for some other thought process, and that it is only a question of revealing this substitute correctly in order to reach the hidden signification of the dream'.[3] Freud also concluded that the dream is often the fulfilment of a wish. In this case, the wish that he had acted differently, that he could erase his responsibility for Irma's protracted sickness. Likewise, in my dream, I must have tried to avenge my guilt at failing to keep my commitment, by turning the reproach back upon Esra for being late to the imaginary appointment of the dream.

Still in bed, slowly emerging from a cloud of intense mental rumination, I raised the blinds and looked outside. It was another beautiful day in the eternal city and I had no commitments. I believed that a long walk would do me good, so I set out into the street heading to the centre and the river intending to make the most of the day.

## Bad conduct

Guilt involves misconduct, or even just the belief of having done wrong. And it is generally some wrongdoing that offends, overlooks or causes harm to someone else, often in violation of a rule or a social norm. It entails judging right from wrong, discerning what is acceptable from what is despicable, advantageous from hurtful. An unjustified burst of rage towards someone we care for or an excessively snappy reaction, such as in Bruce's case, makes guilt supervene. Guilt is a moral emotion, perhaps the quintessential moral emotion, and is therefore about values.

When considering complex emotions such as guilt, conceit, vanity or humility, Darwin wondered whether they could be identified clearly and unmistakably by any distinct physical expression, and acknowledged it to be difficult. Some of his foreign correspondents who searched for snapshots of emotions across the world did provide him with a few answers. For guilt, what they mainly referred to was the facial expression of someone who avoided the gaze of their accuser by keeping the eyelids lowered and semi-closed, giving the accuser only 'stolen looks'.[4] Darwin reports having read an expression of guilt on the face of his own two-year-old son, who gave away his unspecified 'little crime' by an 'unnatural brightness in the eyes, and by an odd, affected manner, impossible to describe'.

Why do we feel guilt at all? Where does it come from, and what is its use?

It is more or less intuitively clear why one would benefit from the capacity to feel anger, despite the outflow of energy connected to our uncontained bursts of rage and the ruinous, dangerous forms the emotion may take: anger is a strategy we have developed to defend ourselves from attack. It is a mutinous protest against any violation of the delicate borders that safeguard our survival and, I would say, respect.

Like anger, guilt is shaped by personal values and by the behavioural codes and norms of the culture in which we live. However, guilt is anger's reverse. We feel anger when another person offends us. We feel guilt after we have ourselves offended or violated someone. I can list at least a dozen flavours of this destructive emotion besides the guilt I felt for not keeping a commitment. Just to recall a few, think of the guilt you may carry with you for arriving late to work, or missing a deadline. Then there is the guilt your parents may impose on you if you neglect calling them for more than a week or you have chosen to live thousands of miles away from them. We are capable of inflicting guilt upon ourselves for doing or failing to do something: skipping a yoga lesson, say, and nevertheless ingesting irresistible crisps at the pub, or failing to quit smoking. Forgetting to respond to an email may haunt us for an entire weekend. Guilt assails us when we feel we have neglected or been snappy with our partners or even when we are more successful than they are. It is even possible to feel guilty for being happy!

We also use guilt to manipulate others. We may make employees feel guilty about their mistakes and may similarly make family members feel guilty for demanding too much or giving us too little. I could definitely go on with the list.

On a daily basis, and through the years, the load of guilt adds up interminably and sinks so deep inside us that it becomes hardly possible to eradicate it.

Guilt loads us with fear. Guilt gnaws. It bites. It attacks relentlessly. It's like a pebble in your shoe that you wish you could get rid of, or some heavy burden. A stinging insect. All such common metaphors apply.

However we personally feel the pressure of guilt, it's fairly certain that we spend – or waste – a lot of time ruminating on it. Now imagine a life, your social and interpersonal life, void of *any* kind of guilt. If you haven't already dismissed this as a ridiculous exercise, but are taking seriously the possibility of a guilt-free existence, you are probably thinking: what a relief it would be! In view of all the various instances that can produce, prolong and generate new guilt, we would certainly gain a considerable amount of time and peace of mind.

However, if we did not or could not feel guilt, we would repeatedly make mistakes. There would be no incentive to alter or improve our conduct. We would disregard any form of social and moral norm, overlook the consequences of our actions. Repenting murderers fight with a sense of guilt to the end of their lives. By contrast, psychopaths often don't feel guilty. So, biologically, guilt has evolved as a social reparative tool that ensures certain actions will not occur, or are not repeated. It sculpts a better version of ourselves. It curbs personal interests and makes space for altruistic and pro-social deeds. The feeling of guilt is indeed unpleasant, long-lasting and hard to eradicate, but, that being so, it inspires action to repair the damage done (for example with an apology) and attempts to stop, undo or make up for the consequences of the offence perpetrated. Guilt is, therefore, a strong motivator to act in morally and socially accepted ways and to correct our conduct.

My main aim in this chapter is to tell you what neuroscience has learnt about guilt and where scientists believe it hides in the brain. Before that, I will also tell you how guilt is connected to concepts of moral purity and of the special relationship it entertains with time and memory. But first of all, I am going to briefly introduce you to some of its friends.

## The pang of guilt, the sourness of regret, the heat of shame

Guilt is often misinterpreted and mistaken for other emotions, especially regret or shame. There are similarities between these emotions, but also fundamental differences.

Both guilt and regret entail decisions and choice of actions – or omissions of actions – with often unwanted consequences, but regret is morally less intense. We experience regret when the outcome of our decisions turns out to be less desirable than what we expected, or less favourable than a discarded option. But unlike a guilty action, a regrettable decision does not harm others. For instance, imagine you forget your clothes and shoes in the bathroom after taking a shower. If, later, you stumble upon them yourself and break your arm, you will feel regret, but if it is your little brother who falls and breaks his arm because of your negligence, you will feel guilt.[5] Regret is also the emotion of missed chances. For instance, you may regret all your life having wasted four years of your youth in law school, following the advice and insistence of your parents, realizing only later that law wasn't exactly for you and that mathematics or art would have been a better choice. Or you may regret having postponed, for lack of courage, initiating a conversation with a beautiful passenger once seen in the tube.

Of greater interest is what distinguishes guilt from shame. These two emotions are indeed similar in that they both speak to our moral self. When we are ashamed about something, we shrink, we turn inward. We feel inferior, inadequate, unworthy. We would like to get out of sight, disappear into a hole in the ground. The very moment I feel ashamed for doing something, I actually feel as if a swift fire were consuming me. Shame, too, may lodge itself deep inside in our psyche and leave profound wounds. Shame can be destructive.

Guilt and shame also often co-occur. The friction of guilt ignites the heat of shame. Psychological research has revealed some of the significant but fine differences between the two.[6] One main difference is between the public and private spheres of guilt and shame. While guilt is considered a private and solitary experience, characterized by the rumination over our wrongdoing, shame is intrinsically public, because it originates in exposure to other people's judgement of behaviour, mistakes or transgressions from our past that we consider unacceptable or disgraceful. Basically, guilt happens in private, whereas shame has an audience.[7]

Perhaps the best way to distinguish guilt from shame in another person is to look at their face. Blushing will give shame away. Blushing is part of the physiological responses that come from shame, not guilt. Even if your conscience may sting you, you don't blush because you feel guilty, you blush because of what others might think of your actions. And it is common to be more acutely sensitive to reproach and blame than to praise and admiration. Your cheeks, your neck and sometimes your ears crimson. A general tingling feeling pervades your entire body.[8]

## Wash away your guilt

As I crossed the Sant'Angelo Bridge, I couldn't resist gazing at the beauty of the dome of St Peter's cathedral on the other side of the river Tiber. So perfect and dominant over everything else. So magnificently harmonious and intimidating at the same time. I stood for a couple of minutes enjoying the view, breathing in the bluest of skies, in the unusual quietness of an early morning in the centre of town. Guilt is a deeply pervasive narrative in Christianity, I would say one of its greatest instruments to instigate and shape good moral conduct. Guilt stains us. It makes one feel dirty. It is associated with feelings of impurity. The Church frequently reminds us of our sins and invites us to redeem ourselves, through confession, punishment and the reparation, where possible, of our wrongdoing. Cleansing actions are used to wipe away moral impurities. Baptism is a symbolic cleansing, the water being supposed to wash away even the Original Sin, the one shared with Adam and Eve who plucked an apple from the tree of knowledge.

But whether you are religious or not, if you are aware of your conscience, bad behaviour will make you feel guilty. And if you feel guilty, there is a chance that you will find yourself horrible, even disgusting. Guilt is intricately connected to disgust.

Evolutionarily speaking, the ability to feel disgust has offered the advantage of despising and avoiding rotten food, or food fouled with unwanted contaminants. Disgust is a remonstrating emotion that begs for a return to purity, to the elimination or separation from whatever element has contaminated it. We say, for instance, that we are 'clean' if we haven't taken drugs. We are also 'clean' if there is no pathogen inside our bodies, for instance if a test for viral or bacterial infection proves negative.

Just as this visceral feeling of disgust is a reaction to physical contaminants, the disgust elicited by guilt is revulsion at moral violations, a kind of moral indignation towards thoughts or actions we disagree with and find deplorable. For instance, we may find someone's opinions disgusting. We can feel moral indignation and disgust towards an entire political system or a terrible chapter in human history. Charged and palpable, the emotion of moral disgust has lately marched along the streets of many capitals in the world in protest against the greed and corruption of bankers and politicians, in light of the mishandling of the economic crisis. What all demonstrators shared was a sense of indignation.

In English, as in many languages including Italian, moral integrity is also figuratively expressed through images of purity. For instance, our conscience is 'clean' if we deem our conduct impeccable. If we have never had problems with the law, we have a 'clean' criminal record. In the brain, there is overlap between regions involved in the feeling of visceral disgust at rotten food and regions involved in moral indignation.[9] There has been a study showing that parts of the orbitofrontal cortex were involved when people made decisions about supporting or rejecting charitable organizations which had views different from their own on gun control, death penalty or abortion.[10]

Another original and interesting study investigated the association between morality and physical purity and involved soap bars, stories and antiseptics. First, a group of researchers checked whether people readily thought of physical cleanness when exposed to concepts of moral impurity. Participants were invited to summon from their memories either an ethical or an unethical action and describe the emotions connected to it. Later the same participants were involved in a word game. They were asked to convert sets of letters and spaces into meaningful words by filling the gaps. For instance:

W _ _ H
S H _ _ E R
S _ _ P

Take a moment to think about these fragments. How would *you* fill them?

Well, according to the study the answer would very much depend on the current state of your conscience. It turned out that those who had recalled the unethical action more readily composed the words wash, shower and soap, which obviously have to do with cleansing. By contrast, those whose recalled actions were not unethical filled in the gaps to compose more neutral words such as with, shaker and ship. Next, all participants, regardless of whether the story they recalled was ethical or unethical, were offered a small gift: they could choose either an antiseptic wipe or a pencil. Seventy-five per cent of those who recalled an unethical story went away with the wipe![11]

## Guilt and time

On one of his regular visits to the home of two of his closest Parisian friends, the painter Avigdor Arikha and the poet Anne Atik, the Irish writer Samuel Beckett carried with him a heavy edition of Immanuel Kant's complete works. As Atik remarks in a memoir of their beautiful friendship, ironically sandwiched between the pages of the *Critique of Pure Reason* was a short manuscript of a poem entitled *Petit Sot*, which means Little Fool. The poem dealt with Samuel Beckett's earliest conscious feeling of guilt.[12] As a child of maybe five or six, Beckett had innocently placed a hedgehog in a shoe box. He dearly loved and truly wanted to protect the animal he had found and even fed it daily with worms, but one morning, to his infinite dismay, he discovered it dead. Anne Atik says that, as an adult, Beckett told his friends this story on several occasions. This regrettable episode had haunted him throughout his life and he had never been able to repress it. It touched him so deeply that he felt the need to express it in a poem.

Emotions in general entertain a special relationship with memory. Episodes void of emotional importance are easily forgotten. On the contrary, those laden with strong emotions, positive or negative, grow strong roots. Guilt punctuates our autobiography. It dots it with memories that reach far into remote moments of our past. I still remember several childhood episodes that induced a sense of guilt, even those children's 'little crimes' – as Darwin called them when describing guilt in his son. For instance, I can't forget the time that I whipped away the chair as my sister was sitting down, causing her a somewhat painful landing and a huge bruise, even though it happened long ago. My parents scolded me and punished me for that.

Several studies have investigated the autobiographical recollection of guilt-linked memories. One in particular looked at their distribution across time.[13] Are memories connected to moral actions different from other kinds of emotional memories? In other words, can the burden of blame weighing on an event, an action or an omission of an action influence their memorability?

A team of psychologists elicited moral memories in a group of people by cueing them with words connected to moral feelings or actions, positive as well as negative: for example, 'honest', 'responsible', 'virtuous' and 'compassionate' as well as 'stealing', 'unfaithful', 'cheating' and 'sneaky'. It turned out that their memories of positively moral feelings or actions mostly related to the recent past, while the memories connected to negative moral events were mostly confined to more remote periods in their lives. These results, while they give added evidence that morally heavy actions, including those associated with guilt, can't be easily forgotten and that we are capable of recollecting them even if they took place in the remote past, also raise another interesting point. There is a certain bias in the recollection of morally problematic memories. It seems that we have a tendency to re-create our autobiographies, associating with our recent past mostly actions that make us appear as 'good' people, whereas the negative deeds are pushed back into the remoter past. It is as if we acknowledge the fact that, yes, we have been bad, but we prefer to believe that we are currently a better person than we used to be. A preference to believe that we are improving accords with the idea of moral feelings such as guilt having a reparative role in our lives.

## Choices and more choices

Consider the following dilemma. It's a fresh, sparkling spring Sunday afternoon and you are attending a friend's wedding celebrations in a beautiful house out of town.[14] While everyone else is hovering over the buffet indoors, you decide to take a breath of fresh air and check out the surrounding gardens until the queue for the food recedes. As you are walking around, you notice that in a small shallow stream, a child is about to drown. Desperate, she is waving her hands to demand help, as she struggles to keep her head out of the water. What do you do? Your first impulse is to save the child as fast as you can. You know that you could do this very easily, but you also realize that in doing so you would ruin the new designer suit that you bought for the occasion and that cost you over £2,000.

For almost everyone, there is truly no hesitation. There is absolutely no item of haute couture that is worth the life of a child. It would be a morally terrible, hideous and deplorable act to let the child drown just to preserve a piece of clothing, however precious and elegant that may be. Letting a child die would make you feel guilty for the rest of your life and there is something inherently wrong in it.

Now, consider the following. One evening when you come back home you find a letter from an international charity organization reminding you that, in some parts of Africa, children have no access to drinkable water. By donating a small sum of money – say around a couple of hundred pounds or less – you could easily save the life of at least one of these children. Again, you could rush to pull out your credit card and complete the online form on the charity's website to send the money in the direction of the child in need. But again, you realize that by not making the donation you could put the money towards a trip to Bond Street to buy an Armani suit or other luxuries – unnecessary for your survival – you have always wanted to wear. What would you do in this second case?

Moral philosophers point out that there is no moral difference between the two scenarios. In both cases, at stake is the life of a child. Yet, when confronted with the second set of choices, most people find it acceptable and morally impeccable to put the charity letter aside and ignore the plea to save a child in a remote part of the world. Most people can do that without experiencing a nagging sense of guilt. They might on occasion feel a sense of guilt after a wild shopping session, but usually this doesn't prevent them from doing it again.

The philosopher and neuroscientist Joshua Greene, who has used the above scenarios for his research, argues that the difference between them lies in how closely they touch us emotionally. Discovering the child in danger of drowning directly stimulates our emotions. Our proximity to the child, the immediacy and urgency of the risk of death, the fact that we hear her cry and see her waving her hands, that she is desperately asking for help, all send a direct message to our emotional networks in the brain. By contrast, receiving a letter in the post that tells us about children who are also in danger of dying, but who are far away, does move us, but probably not to the same extent. If we don't donate the money, maybe someone else will.

As we have seen, there is no doubt that emotion affects moral judgement.

We learnt from the colourful stories recounted in the last chapter that damage to the prefrontal cortex, in areas overlapping the orbitofrontal and the ventromedial sections, makes individuals disinhibited and irresponsible, unable to control their social conduct, insensitive to social norms and standards of appropriateness and more prone to violations of values. In some cases, both among those where the damage is due to an incurred injury and those where it arises from developmental abnormality, these individuals can't contain aggression and manifest violent behaviour. Some display sociopathic behaviour and are not capable of feeling remorse. The gambling experiment with the cards showed that emotion guides our actions and decisions.

Greene and his colleagues used brain imaging to understand

how the brain operates when people face dilemmas of this kind. The difference in the degree of 'personal relevance' and 'emotional proximity' showed up in the brain images they collected. Indeed, judgements over situations like the child drowning in the stream engaged brain areas that are associated with emotion, while decisions about situations like sending money to third-world countries did not.

In light of their results, Joshua Greene and others have argued that there is an evolutionary reason why we would hasten to save the child in the stream and put away the donation letter instead. In evolutionary terms, receiving a letter, or an email for that matter, asking one to donate money for a child far away is a modern scenario, facilitated by today's large global networks of communication. Our biological ancestors were more likely to have found themselves in the situation of having to rescue someone who was in danger by putting themselves at risk. Our brains, and in particular the circuits of our brains that mediate emotion, have been trained for thousands of years to respond to moral situations of that kind. By contrast, our reactions to the more distant cry of children in remote places haven't had the reinforcement of years of evolution.[15] The decision to act towards saving their lives involves more sophisticated reasoning.

## The deep seat of guilt

Guilt is central to dilemmas such as the one described above. Not helping the child would be an incredibly heavy burden to carry, whereas not donating the money allows us to comfortably go on with our lives and spend money on luxuries and surplus commodities we don't need, with a lesser sense of guilt.

As I said earlier, guilt is essentially about choices that can directly or indirectly have an impact on others, or violate norms that are agreed upon in a given society, either explicitly, such as in criminal codes, or implicitly, as in customs or conventions.

For a long time, guilt was a scientific subject for psychology, not neuroscience. It was about testing decision-making, attitudes and behaviour in given moral choice scenarios, in individual settings or in simulated social groups. Scientists are now trying to integrate those tests with contemporary brain science. These days, that normally involves using brain-imaging technologies, in particular functional magnetic resonance, or fMRI. A means by which measurements of blood flow in the brain can be captured and translated into images, fMRI has evolved as a key research method to visualize the brain's operations as they take place in real time. This is indeed a daunting task.

Metaphors of guilt's overpowering and long-lasting nature would easily lead us to construe images of guilt occupying a deep seat in our brain, engraved in hidden neural grooves, and constantly pounding, like the pang of an irrepressible bad memory. But if we feel guilty about something, does that mean that some part of our brain will be continuously sparking guilt? After all, despite guilt's incessant effect, we feel it more keenly when we are reminded of our bad deeds.

Studies investigating the neural seat of guilt have consisted in monitoring what happens in the brains of participants in a variety of moral scenarios. In some cases, they were asked to judge hypothetical scripts of social and moral actions, similar to the dilemma discussed above, or to choose whether or not to cause harm to someone. In other experiments participants were exposed to emotionally charged scenes representing social violations, such as physical assaults, while in yet others they simply read or listened to guilt-laden sentences.[16]

Ullrich Wagner and colleagues at the Charité Institute in Berlin, Germany, conducted a different kind of study. The singularity of their experiment was the exploration of the neural seat of a personal, self-conscious sense of guilt, the one that germinates in the remembrance of guilt-associated events, like Beckett's pungent memory of the accidental killing of the hedgehog.[17] Another particular element in this study is that it aimed at mapping the brain's specific nook for guilt, by comparing what happened in the brain during the recollection of guilt with what happened in the brain during the recollection of shame, guilt's false friend, and sadness, a less related emotion. To do that, they asked over a dozen people to first specify in a list events from their past (since the age of sixteen) marked by a deep and powerful private sense of guilt, as well as by the other two emotions.

Without mentioning by name the actual emotions in question, the team of scientists sought to obtain from the participants descriptions that for instance involved the transgression of rules or damage to others in the case of guilt, situations that jeopardized personal honour or reputation in the case of shame, or themes of loss in the case of sadness. This way, the entries of all participants for each emotion would share basic commonalities but would be free of bias arising from each individual's personal definition or conception of those three emotions. For each event on their list, the participants then also provided keywords that were supposed to trigger recall of that event. Someone who had cheated in a history exam, say, might have given 'history' as their keyword, but they could also have said 'rain' if it had been raining during the episode they described. During the scanning procedure, people were prompted with the memory-laden keywords and asked to try to relive the emotion experienced during the guilt-stained event. A similar procedure was used for the other two emotions.

As you would expect, since the experiment involved evoking memories, when Wagner and his colleagues analysed the brain-imaging data they noticed activity in areas of the brain participating in memory retrieval. But the imaging results also pointed to areas in the anterior part of the brain, in the prefrontal cortex. Roughly speaking, part of the orbitofrontal cortex and parts of the dorsal medial prefrontal cortex were engaged during the elicitation of guilt, but, importantly, not during the recollection of shame and sadness (Fig. 4). From what we have learnt about these two regions in the prefrontal cortex, these results are not surprising. Since guilt has to do with choice and moral decision-making, we would expect it to be at work in brain areas that are in general involved in inhibitory control of behaviour, which is necessary when we calculate the consequences of wrongdoing or causing harm.[18]

But can a brain scan indeed convey a deep sense of guilt? And

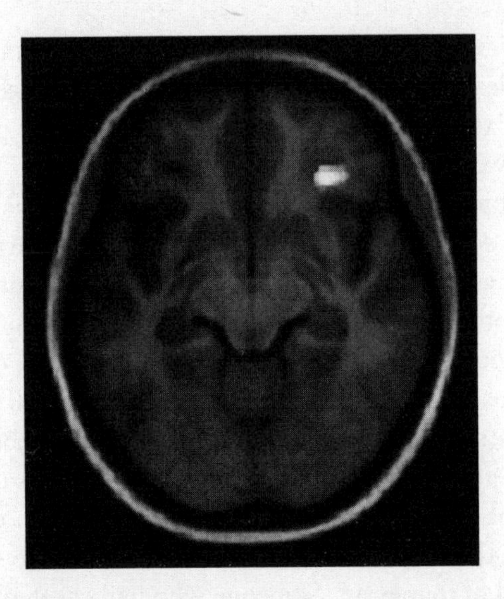

Fig. 4 Brain activation for guilt. From Wagner *et al.*, 2011, Cerebral Cortex, by permission of Oxford University Press

what does it mean to have identified regions in the brain that 'light up' when guilt is recalled?

It would be hazardous to claim that by the means of brain imaging we have narrowly mapped the deep seat of guilt, let alone that a particular region is responsible specifically for the feeling of guilt and not, for instance, shame or regret.

The image of a brain scan that is supposed to have trapped guilt in the brain is not particularly helpful either in understanding why it is so hard to get rid of a nagging sense of guilt, still harder to assuage it.

But while I was in Rome, I gained a better grasp of the meaning of guilt from another image, a timeless painting in a museum.

## A restless genius

From Piazza del Popolo, I climbed the many steps of the Pincian Hill. Before my trip to Rome, a sculptor friend of mine, who had a passion for the painter Caravaggio and had developed an interest

Fig. 5 Caravaggio, *David with the Head of Goliath* © Alinari Archives/CORBIS

in guilt, suggested I go to see some of the master's paintings at the Galleria Borghese. In particular, he recommended I should look at *David with the Head of Goliath*, a canvas depicting the biblical story of David's triumph over the Philistine giant Goliath (Fig. 5), which hangs in a relatively small room packed with many other works.

After a long queue outside, I finally made my entry into the building and was happily thrown back in time among extraordinary pieces of Renaissance and Baroque art. Tourists swarmed in the hot rooms, pacing the magnificent marbled floors and walking around statues. When I reached my intended destination, a small crowd was gathered around the painting, so I waited until it vanished and I could stand in front of the picture by myself. The view is difficult to erase from one's mind. It is a dark, intensely penetrating picture you sense is hiding something sinister. Caravaggio's renowned mastery of chiaroscuro – that is, the sharp contrast between light and dark – works perfectly here. A sombre meaning emanates from every inch of the canvas. A severed head still dripping blood swings by the hair from the hand of David, who holds the gleaming sword with which he perpetrated the decapitating blow.

Art is extremely powerful at summoning emotions and at instigating a dialogue between an object and its viewer.[19] The effect on me of that viewing was immediate. I was enraptured by it and found it resonated with some of the difficult thoughts I had entertained that morning. This became all the more evident after I learnt more about the circumstances of its creation and the life of this extraordinary master of painting.

Born in Milan and raised in a small town called Caravaggio, Michelangelo Merisi (1571–1610) – who later was simply named after his village of origin – arrived in Rome when he was about twenty, keen to find success and the appropriate milieu in which to develop his talent as an artist. Within a few years, he became the most famous painter in the city.[20]

Caravaggio was definitely not an easy-going chap. He was arrogant, uncompromising, irritable and touchy. No stranger to the courtroom, Caravaggio had a criminal record that rivalled his artistic achievements, for it seemed that when he wasn't painting he was getting himself into one brawl after another. During his life in Rome he was accused of harassing women, messing with guards, attacking waiters – he once threw a plate of artichokes at one. He was also put on trial for libel.

The painting of David and Goliath originates from a crucial specific episode in Caravaggio's life. On the night of Sunday, 28 May 1606, at the age of thirty-five, this genius of the Roman artistic world became involved in a sword fight that culminated in his opponent's death and left him a hidden fugitive for the rest of his life.

A capital sentence – a *bando capitale* – was imposed upon Caravaggio as the murderer. This sentence meant that anyone who found him was entitled to report him to the authorities or even kill him and deliver his head – his *caput*.

While away, Caravaggio never ceased longing for a return to the bustle of the city of Rome. During this period, he also painted incessantly. The exile was one of the darkest and hardest phases of his existence. In spite of that, or indeed because of his gloomy desolation, he created some of his most expressive images, among them the painting I stood in front of.

A very important detail about the image must be revealed. Before Caravaggio, several artists had painted themselves as David. Caravaggio's version of this celebrated scene of good victorious over evil is unique in that it is the severed head of Goliath that is Caravaggio's self-portrait. In Caravaggio's painting, David bears a candid appearance and shows no exultation in his victory, but rather expresses compassion and pity. Caravaggio's face is tormented and heavily disfigured by death.

By serving his severed head to the viewer, Caravaggio is expressing his repentance for his actions and attempting to assuage his sense of guilt.

On David's sword, on the side of the hilt, is an acronym, barely readable unless you move close to the painting: H. OC. S. These letters stand for the Latin words *humilitas occidit superbiam*, that is: humility kills pride. It is supposed to be a sentence taken from St Augustine's reflection on Psalm 33 in which he compares David's victory over Goliath to Christ's triumph over the devil.[21] Good prevails over evil. In one painting we have a whole host of moral emotions. Guilt, backed up by humility, promises to restore good conduct.

## The truth of context

It is entirely disputable whether Caravaggio in truth felt any guilt. There is no way to find out. In light of his turbulent past of crime, brawls and violence, he may well have felt none. The fact that he used his own face to depict Goliath is no definite proof of his feelings of remorse. There are no documents or letters that may testify to an authentic repentance. Some argue that his portrait as Goliath is yet another expression of his narcissism.[22] The painting may have been just the artist's nifty stratagem to regain credibility and have the gates of Rome opened to him again. Caravaggio had the painting sent to a powerful patron in Rome, the Cardinal Scipione Borghese, the administrator in chief of the Vatican system of justice, to seek forgiveness and permission to re-enter the city from which he had fled in disgrace.[23] Caravaggio's undeniable talent, his boundless imagination and his sensitivity may have enthralled anyone willing to give him another chance. If his goal was to convey a deep sense of guilt and his repentance, he did succeed in it. He certainly knew how to conquer the viewer's sympathy with the emotional power of his paintings.

We need to pay attention to the historical context of the painter's life. In Caravaggio's Rome, murders were not a rare occurrence. The prevailing customs and squalor of the city were such that fights, or even homicides, happened on a regular basis. Rome was a daily circus, a rowdy and perilous place. This does not mean that in Rome at the turn of the seventeenth century murders were encouraged, or that they would go unpunished. But they were frequent. The anatomical precision and realistic immediacy of the physical violence in Caravaggio's paintings reflected first-hand knowledge of the violence to which he was exposed on the streets.

What makes emotions such as guilt and shame moral is also their dependency on given values of the social context. As a moral emotion, guilt is influenced by the behavioural codes and norms of the culture in which it is experienced. Actions or turns of speech that are considered inappropriate in one culture are guilt-free in another culture. In the UK, homosexuality was not decriminalized until 1967. For almost all religions, it still remains an unacceptable sin, and several countries in the world, such as Uganda or the United Arab Emirates, continue to ban it.

Today, killing someone would never be regarded as an acceptable custom or a forgivable deed (though, that said, there are countries that inexplicably retain the death penalty). However, when judging the severity of a murder, courts take into account elements that may justify the killing – say, legitimate defence. In countries such as Italy, crimes of honour were customarily punished with lenient sentences until the early 1980s. If an action is not frowned upon or considered illegal in a particular society or social context, those committing it experience no habitual response of guilt. The biological apparatus that can make us feel guilt is spared the expenditure of energy. So, morals and norms evolve and change in society, and our biological ability to make moral choices and feel guilt over them adapts accordingly.

Caravaggio eventually received the Vatican's pardon, but he never reached Rome, for he died in mysterious circumstances on his way back to Rome.

If Caravaggio were still alive today, he would certainly make a very interesting subject for neurological study: both in further investigation of the neural seat of guilt, and a thorough scrutiny of his extensive portfolio of violent and rebellious actions. Was he a carrier of the short version of the MAOA gene? What did his prefrontal cortex look like? Did his solitary childhood and dismantled family play a role in the outcome of his violent behaviour? The answers blow in the wind.

But the incomparable calibre of his art, his enhanced imagination and his capacity to trap a fleeting rainbow of emotions on canvas persuades me that he must have felt unease and discomfort after committing the murder and that guilt cannot have left him unscathed.

## What's in a blob?

To compare a brain scan with a Caravaggio painting in the search for the most authentic representation of guilt may be novel or sound unusual to you. Take a look at both images again. First, examine the blob in the fMRI image, and then gaze at the painting. Both are supposed to represent the emotion of guilt. They are powerful images, each in its own way. The scan is extremely technical and hard to make out, if you are not familiar with brain anatomy. Where is that dot exactly, if you were to imagine it in your own head? The painting is undeniably intense, extremely sombre, but also requires knowledge and interpretation beyond the immediate, communicative strength of its treatment of light. Nevertheless, they both entice a viewer's attention.

Attractive images of the brain, especially scans of someone experiencing guilt or other emotions, abound. Emotions are mediated by brain activity. Just as it is useful to observe the outer appearance of emotions, in facial expression, skin conductance response or body movements, inspecting the brain reveals fundamental components of emotions.

The greatest advantage of functional magnetic resonance imaging is the possibility of watching the brain without having to open the skull. Earlier, in order to inspect the brain's turns and grooves, you had to drill down through the skull or examine the brain outside the body. Now, we can watch what goes on inside while the brain is engaged in all sorts of tasks. More than a snapshot, an fMRI image is a still from a movie. It aims at capturing brain workings in space and time. This is certainly an incredible and unprecedented privilege. However, there is still a problem of refinement.

A detailed, thorough explanation of what happens when you enter the large fMRI scanner would involve going into details of complicated engineering and quantum mechanics. But even without a degree in physics, it is possible to grasp the essential features of this technique and to understand both its power and its limitations.[24]

First of all, it is not entirely correct to say that the colourful blobs that stand out against a grey, blackish background in a brain scan are direct signs of brain activity. The blob signal on an fMRI scan, however narrowly localized it may be, is primarily telling us that there is a lot of oxygen in that area, brought in by the flow of blood, which we assume is needed by neurons to be able to function, just as more blood rushes to the stomach during food digestion for the absorption of nutrients.

Basically, if a part of the brain is needed to accomplish a given mental task – say remembering a seven-digit number, which keeps the prefrontal cortex busy – it will require energy to carry it out. Where does the energy come from? Like muscles, to carry out their work neurons need sugars, such as glucose, which are broken down in the presence of oxygen.[25] The oxygen is rushed to that site through the haemoglobin carried by blood. In effect, what is being detected during an fMRI session is the ratio between the amount of oxygen brought in and the amount of oxygen used up for the task, as signalled by the presence of oxygen on the haemoglobin molecules in that area. The oxygenated and non-oxygenated forms of haemoglobin have different magnetic properties – the protons in their atoms behave differently – and this difference is picked up by the huge magnet of the scanner (the magnetic properties of haemoglobin were discovered back in the 1930s by the great scientist Linus Pauling!).[26] In laboratory parlance, this difference is called BOLD (short for 'blood oxygen level dependent') contrast. So, what the brain scanner picks up are incredibly tiny differences at the subatomic level of blood.

As you would expect, glucose and oxygen are needed across the entire brain, including areas that are not engaged in any particular task. There is a great deal of background activity the brain carries out without our realizing. What fMRI does is either to map the location during a specific task of any progressive increase in oxygen relative to levels of oxygen in a control quiet state (also called the *baseline*, or default state, where the brain under scrutiny is at rest), or to measure the difference between changes in oxygen during two different tasks. fMRI is basically looking for and detecting alterations, the additional activity that is associated with the task. So, for instance, in the experiment looking for the seat of private conscious guilt, the signal detected showed alterations in activity between moments of guilt recollection and a baseline state, as well as differences between recollections of guilt and shame, or guilt and sadness.

## Seeing is believing

Neuroscience holds enormous allure for the non-expert public. A study found that the same neuroscience result was regarded by non-experts as more credible if it was represented by the image of a brain scan than if it was presented with a more traditional bar graph or no image at all.[27]

Seeing is believing. Perhaps brain scans are more persuasive because they offer a physical explanation. They are seductive in that they increase the plausibility in the eyes of the general public of conclusions presented by researchers. They have become icons comparable to X-rays and the DNA double helix in their importance and resonance in our culture. They are found on the covers of books about the brain, in tube ads, in promotional literature for corporate and managerial courses that are supposed to improve performance.[28]

As Susan Fitzpatrick reports, in 2005, at a meeting of the American Association for the Advancement of Science, a panel session organized by the James S. McDonnell Foundation was given the provocative title: 'Functional Brain Imaging and the Cognitive Paparazzi: Viewing Snapshots of Mental Life Out of Context'.[29] The analogy between the work of brain-imaging scientists and that of aggressive photographers avid for a scoop may sound at first unusual or slightly off-beam, but it does have a point. What paparazzi do is to steal intimate, private moments of celebrities and then publish them on tabloid front pages. The raw pictures are 'repackaged', revisited and extracted out of their wider original context so that, with the help of juicy misleading headlines, the snapshots of an occasional long face, a solitary walk in the park or unusual weight-loss are sold as evident signs of depression, an imminent divorce or hidden eating disorders. Through powerful magnetic field scanners, cognitive scientists also capture private instants of our mental life. They certainly have no intention of gossiping, nor of falsifying their data, but again, the resulting pictures are mere extracts of the mind. When fMRI images hit the newspaper headlines, they too are taken out of context, out of the laboratory setting where they were produced.

I mentioned earlier that fMRI imaging attempts to capture the workings of the brain in space and time. Some remarks about scale need to be made in this regard.

Time is a critical issue. The speed of blood flow in the brain is measured on a scale of seconds, whereas the subtleties of neuronal activity are hundreds of times faster. So there will always be a certain time-incongruence in the correspondence between neuronal activity and blood-flow dynamics.

Each fMRI image is a colourful, computer-generated map resulting from the comparison of signal intensities across the various regions of the brain. Each little dot composing the image is called a voxel – which is more or less like the pixels that make up your iPad photos, but is a three-dimensional unit of volume, rather than a two-dimensional unit of flat space. What you see is a shade of colour, a tiny pinned spot in the human brain, but behind it lies a vast complex of neural tissue and neurochemical reactions. Each voxel corresponds to approximately 55 cubic millimetres. This equates to around five million neurons, with anything between twenty-two billion and fifty-five billion synapses, which are the points of connection between neurons. If stretched out, the distance covered by the ramifications of the neurons involved would roughly correspond to the distance between, say, London and Manchester.[30]

Scientists perform complicated statistical and numerical operations across that vast territory of neurons captured in the scan. Each voxel is compared with all the others, in search of meaningful information. The shadings in colour intensity are a reflection of the statistical significance of the measured differences. The more intense the colour, the more significant the change in haemoglobin oxygen detected. In 2012, a rather bizarre study showed some of the dangers of such statistical comparisons in fMRI. For its originality and improbability, it even won the IgNobel Prize for Neuroscience – the annual counterpart of the real Nobel Prize, which praises discoveries that 'first make people laugh, and then make them think'.[31]

The humour is inevitable because of the study's unique feature: its only participant was a dead salmon.[32] The researchers placed it in an fMRI scanner and showed it images of individuals engaged in a variety of emotional scenarios. The salmon was then asked to report which emotions each individual was experiencing – I confess I would have loved to be there when the experiment took place. As expected, they received no answer from the salmon, but the researchers identified neural 'activity' in the salmon's brain and spinal cord! How was this possible? The authors remark that as thousands of comparisons between one voxel and the others are made during each fMRI data analysis, there is a high chance that false positives will emerge. You may see stuff that you are actually not supposed to see. There is absolutely no way that a dead salmon could have recognized emotions.

There are methods in statistics that help 'correct' for such mistakes. In fact, when the authors applied those correction methods, the blob in the salmon's brain vanished from their analysis. The dead salmon researchers reported that while such methods are available as part of most fMRI analysis software packages, not every research team applies them, because correcting for false positives may reduce the statistical power of their analysis. They found that, for instance, in 2008, the correction methods had been used in only 61.8 per cent of the papers appearing in the *Journal of Cognitive Neuroscience*, one of the many journals in which fMRI results are published. So, absurd as it was, the salmon experiment highlighted a frequent methodological negligence in the production of brain-imaging data.[33]

Let's go back to Caravaggio's solemn painting. Although the picture taken on its own suggests something sinister, knowing about the artist's turbulent past, or the murder he had committed, would help you discern the presence of guilt. You would indeed probably need that background information. As Darwin noted, guilt is not the easiest emotion to read on a face. He thought we could detect complex emotions like guilt with our eyes, but when doing so 'we are often guided in a much greater degree than we suppose by our previous knowledge of the persons or circumstances'.[34] As usual, the great naturalist had a point. When you look at brain scans, unless you know what you are supposed to be seeing and are familiar with the nature of the study, the intensity of the light and its position are rather meaningless.

Guilt has many shadings, and there are very many different scenarios and behavioural conditions in which it could be measured. The question remains as to whether all these different types of guilt are processed in the same location and by similar processes. This is why if you compare brain scans originating from several separate studies you will find the results differ, sometimes only marginally, but on occasion to a noticeable degree. This brings me to a general observation about the measurement of emotions in fMRI.

When your emotions are being measured inside a scanner, you are often asked to perform a distinct task, for example watching images, remembering events or, as we have seen in studies of morality, making ethical choices such as whether to save a child's life. However realistic these tasks may be in their approach, they can only ever be experimental reproductions of situations that are much more complex, but also more direct and urgent, when they happen in real life. The tasks in the scanner are convenient substitutes for authentic fragments of life. There remains a gap between the two and so far we don't know what the brain activity of the real version of the emotion looks like. Moreover, the blob in the brain scan is actually a representation of the *average* result computed from measurements taken in dozens of individuals recruited for a study. The final image you see is not the oxygen flow of one brain, but the statistically significant oxygen flow across all the participants in the study. Yet guilt works at the level of the individual. It is such a personal, uniquely private emotion that it is hard to imagine it diluted with the guilt of others. Not everyone feels guilt with the same intensity. There are individuals who, without being psychopaths, are simply less prone to guilt.

Finally, there is another analogy I like using when I try to describe what we are effectively seeing when we gaze at an fMRI scan. I think it's like being at the top of the Empire State Building and having a 360-degree night view of the illuminated New York skyline without binoculars. We can appreciate the contour of Manhattan, see more or less where Queens ends and Brooklyn begins, point at New Jersey across the Hudson. Our eyes can perhaps follow the dividing line of Broadway and the rushing trail of cars and spot the dark hole of Central Park. We see the lights of New York life flickering on and off in different areas of the city at different times, and can identify its busiest periods and when things are quieter. Given a pair of attentive eyes and good knowledge of the streetmap we could pinpoint the origin of the glow. We might recognize the lights as coming from a loft somewhere between Canal Street and Washington Square, or up in Harlem and the Upper East Side.

But what we can't see is what actually goes on inside the buildings, the lives and motivations of the people turning on those lights and giving colour and movement to the city. We have no idea if the light is a lamp, a candle or a chandelier or whether it is coming from a bedroom, a kitchen or a living room. We also don't know who turned it on and why: it might be illuminating an intimate dinner or a party or a serious family conversation; it might be on because a child is afraid to sleep in the dark, or because somebody simply forgot to turn it off. So, from on top of the Empire State Building and through an fMRI scan alike, the view is spectacular, but not fully revealing. Right now, the view achieved through an fMRI is crude and approximate. With time, the technique will be refined, giving greater precision and detail on a smaller scale, allowing its potential to be realized.

In sum, when you read expressions like 'this region of the brain *lights up* when you feel fear' (or anger or any other emotion), it is only popular, overused parlance aiming to simplify the complicated underpinnings of magnetic resonance. For me at least, there's no way gazing at an fMRI image can help draw definite conclusions about the sense of guilt, nor map its exact locus, let alone find out how to assuage it.

## The moral brain

At the beginning of the chapter, I explained how guilt appeared to me in disguise in a dream and how the inspiration of Freud's theory on the interpretation of dreams came to him the morning after having had a guilt-themed dream himself, the dream about his patient Irma and the injection.

Psychoanalysis enthusiasts have for a long time tried to find confirmation of Freud's ideas in current neuroscience research. They point out that present-day studies of lesions and modern visualization techniques are drawing a map of the brain that approximately coincides with Freud's structural theory of the mind.[35] The id, the ego and the superego postulated by the Viennese physician are finding their neuro-anatomical seats. By comparing Freud's diagram with today's collected data on the role of brain regions, psychoanalysts have sketched the following general map (Fig. 6). As previously thought, the id would comprise the most internal parts of the brain, such as the stem areas and the limbic system. The ego would lie in the most dorsal part of the prefrontal cortex, and in the rest of the somato-sensory cortex, areas which provide a sense of self and enable perception of the outer world. The ventromedial frontal part of the brain – which overlaps with the areas mapped by the imaging studies on guilt – corresponds to Freud's concept of the superego, the moral apparatus that constrains and prohibits the most instinctual drives. Within this framework, it is

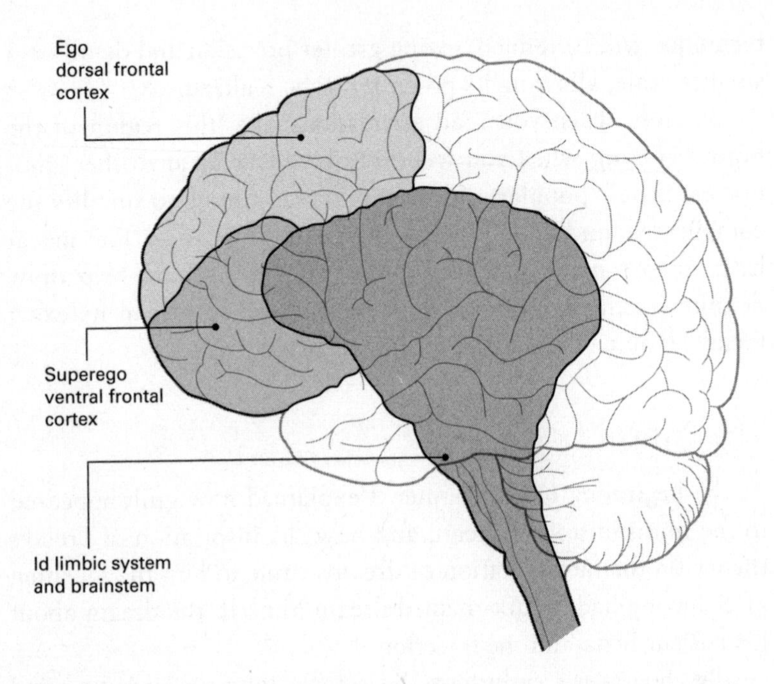


Fig. 6 A brain-map view of Freud's structure of the mind (diagram adapted from Solms, 2004)

not surprising that guilt, as a moral sentinel guarding against or preventing inappropriate conduct, would sit somewhere in areas overlapping the orbitofrontal cortex.

In the past ten years or so, the number of brain-imaging studies attempting to address the neural basis of morality or moral emotions has been impressive. From regret to guilt and shame, a whole list of moral emotions and concepts have been under scrutiny in a brain scanner. Even social comparison emotions such as envy and *Schadenfreude* – the former being displeasure at someone else's good fortune, the latter the relief or joy we feel when the envied person falls from grace – have been investigated.[36]

A few moral psychologists have proposed the idea that all human beings share a basic, universal sense of morality – a concept reminiscent of philosopher Immanuel Kant's 'innate morality' –

and that the brain may even be the seat of a 'moral organ' that helps us choose what is right and what is wrong relying upon unconscious intuitions.[37] The question arises: is morality something ingrained in our biological constitution, or is it something that manifests in society as a direct consequence of behaviour patterns that demand some sort of regulation or norm? Do values have their origin in the brain? Brain-imaging studies of emotions as complex as guilt and concepts as multifaceted as morality are definitely exciting, but in most cases only explorative. What does it actually mean that guilt sits in the orbitofrontal cortex or overlaps with the ventromedial PFC? Are regret and guilt similar because of overlapping fMRI data?

What imaging studies do is delimit by trial and error the area engaged in that emotion.

There is another issue to raise. The idea that the brain works by the operation of distinct modules, each in place for the execution or adjustment of a particular function, is irresistible, but not consistent with what we know of its actual *modus operandi*. From the moment the first connections were made between certain areas of brain tissue and function – for example the discovery of the language region – it was assumed that more and more specialized regions would be identified. But even though the brain displays a fair degree of specialization, the way the brain works is by the integration of connecting pathways and their interactive nature. The case of the emotions is no exception. One region may play a part in several emotions and the neural activity related to each emotion is spread across several regions. Research is moving towards the discovery of *networks* for emotions, consisting of regions working in parallel. One region is, say, specialized for or more strongly involved in one emotion, but plays simultaneously a less prominent role in other emotions.[38]

Over time, the scale and definition of brain imaging will improve. We will also come to redefine and improve the way we observe and measure guilt. For now, we need to accept and take for granted that the exact, confined location of guilt, or any other emotion, is still an estimate: how good an estimate being dependent on the sophistication of current technology and the scientist's knowledge, skill and interpretative judgement.[39]

## Coda

In his *Recipes for Sad Women*, Hector Abad notes with resignation the impossibility of finding dinosaur meat nowadays.[40] He does so because dinosaur meat, together with mammoth's milk, he says, is the only effective remedy to assuage an insistent sense of guilt. It doesn't take long to grasp the irony of such culinary analogy. The chances of getting hold of that prehistoric flesh are so remote that the possibility of assuaging guilt fades as soon as it has been glimpsed.

Abad offers an alternative. Another remedy for guilt is the flesh of a coelacanth, a very rare fish that everyone had thought extinct since the dinosaur era. He reports having come across one himself while fishing in the Indian Ocean in 1946. After some research, he found out its taxonomic name: Latimeria chalumnae, after Ms Marjorie Latimer from East London, South Africa, who had made the initial discovery eight years earlier. A marinated fillet of this rare fish does wonders in fighting guilt, says Abad, its effect lasting about thirty-eight months. Even just a bite is effective.

Apart from these improbable recipes, there are other ways to try to assuage guilt. In the legal system, as we have seen, the perpetrators of a crime may find their punishment and rehabilitation serve to redeem them from guilt (though this will not be the case for psychopaths). But forgiveness remains perhaps the best antidote to guilt: the forgiveness we receive from others, and the forgiveness we may afford to grant ourselves. Caravaggio painted his request for forgiveness. After my museum visit, having absorbed the meaning of Caravaggio's life and painting, I rushed to my room. I had been extremely apologetic when cancelling my acceptance of Esra's invitation, but I felt something had been left undone. So, I decided to write a new letter to her. I reckoned it was the best and only way for me to try to make reparation for my act of negligence and attain some form of forgiveness for myself. 'There is a luxury in self-reproach,' wrote Oscar Wilde in *The Picture of Dorian Gray*, in which he makes the protagonist write a letter to his lover in search of forgiveness for his cruelly abandoning her after she had given a very bad performance as Juliet. 'When we blame ourselves we feel that no one else has a right to blame us.'[41]

I was completely engrossed in the act of writing that letter. I wasn't trying to escape judgement or to sweep my guilt under the carpet. Neither did I expect my feeling of guilt to recede fully. I was simply looking for understanding. Instead of letting my guilt sit inside me, it made sense to give space to it in words. I offered again my apologies and explained my reasons as best as I could. I did feel better afterwards.

After filling page upon page, I went out for dinner. It was the last night of my trip. No coelacanth at hand, but very good wine. Then at midnight, I went to bed, in hopes not to face another bad dream. I was ready for a return to London.

# 3

# Anxiety: Fear of the Unknown

Anxiety is the interest paid on trouble before it is due

WILLIAM RALPH INGE

Anxiety is the handmaiden of creativity

T. S. ELIOT

Just as I was dozing off, the phone began to ring. I had spent a long day in the lab crushing dozens of mouse brains to obtain a few precious milligrams of purified protein, and I had just gone to bed. Exhausted, I picked up after four rings. At the other end was Robert, an old university friend.

'Have you heard?' he asked.

'About what?'

'The world economy is going down the drain.'

'And you called to tell me that?' I yawned.

'It's truly bad this time, believe me.'

It was a cold, dark night in December. Worldwide, stockmarkets slumped, while the number of jobs shed continued to rise. It had been one of the worst days for the economy that year, and I had spent it isolated in a biochemistry room.

Now awake, I jumped to my desk to check the news on my laptop.

'You don't seem to understand.' I could hear the tension in his voice.

'Are you worried?'

'Worried? I'm terrified. I can't even sleep.'

Scanning the headlines, I could see that things were bad. And, yes, I knew Robert had just started working for a major investment bank in the city, one of those financial giants which only one year before had seemed entirely immune to any economic downturn. Things were still going well for him, but he made it sound as if he were barely a few weeks away from being a beggar at the tube station.

'I could always earn money busking,' he said, 'or try once and for all to become a rock star.'

'Robert, I'm really tired,' was all that I managed to say.

'Come on, you work in a neuroscience lab, aren't you supposed to know what to do in these circumstances?' Robert insisted.

'Fix the economic crisis? You're the banker.'

'No, help me cope with anxiety,' Robert replied.

Promising to visit Robert the next day, I ended the conversation, switched off the lights and fell back into bed. But sleep eluded me. Oddly, obscure figures and indices of the economic crisis continued to occupy my mind, like the thought of maths homework left unfinished or a nagging irresolvable equation. My eyes stayed wide open and even though I had a good job and no savings to speak of in danger of evaporating in a cloud of smoke, I found myself worried about the incumbent recession. From there, thoughts roamed freely and became galling concerns. One worry was creating another, for in a matter of minutes I found myself worrying about almost everything. I heard my heart accelerate, my head and chest felt heavy, my throat closed and the following thoughts and questions began to ramble in my mind in disorderly succession:

– Had I switched off the centrifuge properly?

– A rare chronic disease was what kept causing those terrible headaches in the morning.

– Was the front door locked?

– I should not have read that Facebook post.

– What if my university ran out of research funds?

– I was never going to finish the experiments for my next paper in time, so my competitors were sure to scoop me.

– My neighbour hadn't greeted me that morning. Had the party at the weekend been too loud?

– A new red spot on my left arm was the beginning of cancer.

– I still have to buy all my Christmas presents and I won't make it in time.

– The boiler would undoubtedly break down again next week.

– I might never be in a position to buy my own property.

– No pension for me in this lifetime.

– What if I had a bike accident tomorrow?

– Was a new terror attack looming in the distance?

The list could easily go on. Everything seemed to be accelerating towards a catastrophic end.

If examined carefully, some of those worries sound ridiculous, or unnecessary to say the least, don't they? Yet, alone in the darkness of my bedroom, I didn't seem to have much control over them.

Eventually, my worries became something else. Spinning in a vortex of confusion, I began to feel directionless, pondering my whole existence. Just over the threshold of thirty, single, overworked, on the verge of making a leap in my career, I began to worry about the meaning of all I had done, whether or not I had taken the right decisions in life. It was one of those moments when I thought I needed to do everything at once, as if the world were about to end and I only had a few hours left to accomplish all I had ever wanted to do. It felt as if someone had turned off the customary soundtrack to my day, and a strong, stubborn wind had dislodged me from the life carousel I was a part of, uprooting the pillars of hope for the future and leaving an empty stage, with me at its centre, in the spotlight.

That wind had a name – ANXIETY – and it was blowing strong and determined.

Fig. 7 Edward Hopper, *Nighthawks* © CORBIS

When I turned the light back on I was astonished to see that it was still only midnight. I decided to call Robert back.

'Are you still up?' I asked.

'Yes.'

'OK. Meet me for a drink in half an hour.'

So, there we were, a scientist and a banker trying to tame their anxieties in an all-night bar in the early hours of a winter night (Fig. 7).

The circumstances reminded me of W. H. Auden's poem *The Age of Anxiety*, in which four characters discuss their lives, and share their hopes and distress over the human condition in a bar on Third Avenue in New York City.

'When the historical process breaks down . . . when necessity is associated with horror and freedom with boredom, then it looks good to the bar business,' begins the poem.[1] Well, a glass of wine can indeed be of help if times are hard and you are trying to calm down. The characters are: Quant, a clerk, Malin, a medical officer in the Canadian Air Force, Rosetta, a department-store buyer, and Emble, a young man who has recently enlisted in the navy. The mood of the poem is that of uncertainty. The four protagonists feel lost, without a clear direction. Auden began that poem in July 1944, against the backdrop of a war which had left humanity doubtful about the future and hungry for peace. Everyone, he wrote, was 'reduced to the anxious status of a shady character or a displaced person'.[2] Auden was thirty-seven and considered himself 'still too young to have any sure sense of direction'.[3]

Nearly seventy years later, do we still live in an age of anxiety?

Certainly, neither Robert nor I was alone that night. Our anxieties echoed those of millions around the world. The risk of a global recession proved indeed to be real. Five years into it, we are not yet nearing a full recovery. On a weekly basis we hear of terrible news about the general economy and we are all waiting for a resolution that does not seem to arrive. The euro has been on the verge of collapsing several times, with debtor countries like Greece, Italy or Spain at risk of having to leave the monetary union. Our money and the future of our national economies are in the hands of a few suits whom we are asked to trust. The current overall grim state of the economy has affected the well-being and calm of the global population. In the last few years, a daily news diet of layoffs, bankruptcies, fluctuating indices and currency spreads and other financial disasters has caused a worldwide increase in the number of people displaying the symptoms of anxiety, ranging from changes in sleeping patterns, to general nervousness and painful headaches.

In 2010, a report revealed that 52 per cent of people who had lost their jobs to the recession manifested symptoms of anxiety, and 71 per cent reported being depressed.[4] The most affected were those in the age group eighteen to thirty. In Britain, the NHS estimates that one in twenty adults is affected by anxiety.[5] In the United States, every year about 18 per cent of the population suffers from an anxiety disorder.[6] In 2009 the UK government offered psychological help to the millions of people who were confronted with unemployment and debt by increasing the number of therapists and counsellors across a wide network of services that included psychotherapy centres and help hotlines.[7] Anxiety is also a burden for the wider economy. Currently, the annual cost of anxiety disorders in Europe amounts to €77.4 billion, a figure big enough to trigger anxiety itself, and prompting many to consider taking immediate action to fix the crisis and attend to this enormous public health challenge.[8]

Recession aside, we inhabit a world where there is no dearth of reasons to worry. These are both private and global, immediate and remote.

On one hand, we all face the daily pressure of keeping up with work demands, sustaining wide and fierce competition, achieving success and climbing the career ladder. We need to keep on top of our finances, make ends meet on a monthly basis, think ahead and save for the future. We may also be responsible for family, or have to provide support for children, and we are expected to initiate and cultivate social relationships.

On the other hand, the global situation is altogether not reassuring. The world's attitude towards the peril of international terrorism has profoundly altered in the wake of 9/11 and subsequent al-Qaeda attacks, with the military forces of several Western countries mobilized in two major conflicts over the past decade. We live under the constant threat that delicate political and ideological disputes over the construction and use of nuclear programmes in the Middle East may end up sparking a third world war. Unrelenting epidemics – for instance, that of HIV – and the outbreak of new, unexpected fast-spreading infections, such as avian flu and swine fever, are a reality we must learn to come to terms with and that continues to menace the health of the world population.

As if all this weren't enough, we are told that the looming threat of global climate change may indeed irreversibly transform planet earth and initiate major natural disasters. Hurricane Sandy which hit the US east coast in November 2012 may have been a proof of that.

Undoubtedly, each historical period has endured its own

share of different, but equally worrying and serious threats. Biologically, the mechanisms with which we are equipped to counteract such threats and experience anxiety are no different from those of our ancestors. But the frequency and speed at which we are bombarded with news of risks, danger and actual disasters poses an unprecedented challenge to our minds. Turning on the radio or reading a newspaper is enough to be overwhelmed by the load of disquieting events.

As Robert and I sat there talking and drinking, I realized that I had rarely challenged myself to harvest all those hours spent in a brain laboratory to counter a real-life necessity. Each time I told new acquaintances I worked in a laboratory devoted to the study of fear and anxiety, everyone would volunteer to be a subject for one of my experiments, claiming to be material of first quality, the best specimens for research into these dreadful emotions. Yet the meaning of my experiments had all too often remained abstract, confined behind laboratory walls. Talk of brain regions, genes, neurotransmitters and behavioural measurements sounded unbelievably distant from the monologue of personal anxious turmoil. So it was time to understand whether knowledge gathered in the lab could come in handy in such circumstances.

## Fear or anxiety: know your enemy

If you are to defend yourself from your enemies, or to defeat them, you need to know them well. A good first step is to distinguish anxiety from fear.

Fear is one of our basic emotions and by far the most widely investigated in the laboratory. It is classically defined as a response to an imminent threat or danger. When we have fear, it is usually of something specific, of a lion, say, or snakes, or of flying. Evolutionarily speaking, fear is a useful, protective trait that is critical for our survival. It sharpens our senses and prepares our bodies to face sudden perils. If we weren't capable of experiencing it, we would be dead, simply because we would not avoid dangerous and potentially life-threatening situations.[9] Fear makes us swim fast to the shore if we catch sight of a shark, but it dissipates as soon as the shark is no longer a threat.

As so often, Charles Darwin can be of help here. As part of his section on fear in his book, Darwin writes: 'Fear is often preceded by astonishment, and is so far akin to it, that both lead to the senses of sight and hearing being instantly aroused . . . The frightened man at first stands like a statue motionless and breathless, or crouches down as if instinctively to escape observation . . . ' 'The heart beats quickly and violently, so that it palpitates or knocks against the ribs . . . the skin instantly becomes pale, as during incipient faintness . . . perspiration immediately exudes from it . . . '[10] Moreover, the pupils dilate. The guts churn and stir. Breathing becomes shallow. Sometimes, even, hair stands up! Darwin also added that 'terror', by which he meant a state of heightened fear, involves 'trembling of the vocal organs and body'.[11]

This entire set of fear responses occurs unconsciously and within milliseconds. As they unfold, we gradually become aware of them, but we don't actually need to be conscious of them for them to take place. The American psychologist William James makes this clear in his seminal essay 'What is an emotion?', published in 1884. In this essay, James formulated his influential thoughts on how we 'emote'. At that time, the prevailing theory on emotions described them as some sort of mental state of awareness of our reaction to a fact or a change in the environment. In turn, this mental perception would trigger a cascade of physical responses. So, applying this theory in the case of fear: seeing a bear in the woods would first make us be afraid and, consequently, the state of fear would in turn let us start to tremble and shake. James thought this sequence of events was wrong and that what happens is exactly the reverse. He said that we feel afraid because we tremble and shake, not the other way around. Emotions are first and foremost our bodily reactions. Then comes the feeling, or the awareness of them.

He was so convinced about this order of sequence in the way

we emote that he went on to say that if we took away from emotions the bodily symptoms, there would be nothing left. Only a cold and neutral 'state of intellectual perception' would remain.

'What kind of emotion of fear would be left, if the feeling neither of quickened heart-beats, the shallow breathing, trembling lips or weakened limbs, goose-flesh, visceral stirrings were not present, it is impossible to think.'[12]

But let's go back to the distinction between fear and anxiety. Fear has a specific target. What about anxiety? Well, anxiety is not as simple. Anxiety is usually a fear of the indefinite, something that we cannot always explain or even locate in space and time. It is unpredictable, and often the anticipation of an unknown or not necessarily incumbent threat. Just as I did the night Robert called me, we feel edgy and jittery about the possibility of negative or catastrophic occurrences that may never actually materialize. In other words, anxiety is fear that is looking for a reason.[13]

## Anxiety's pedigree

Anxiety's obscure and opaque reasons are good at hiding, but it is worth searching for them. Sigmund Freud devoted a lot of time to this hunt. Freud was convinced that 'anxiety was a nodal point at which the most various and important [psychological] questions converge, a riddle whose solution would be bound to throw a flood of light on our whole mental existence'.[14] Towards the end of the nineteenth and the beginning of the twentieth century, a disease began to seep through modern cities, especially among the upper class and working professionals. It mainly consisted of stomach unease, headaches, neuralgia and general fatigue and was rapidly spreading, much as flu might, in response to the rapid urbanization and the increasingly frenetic and hectic lifestyle spawned by the industrialized world. Across the pond, the American physician George Beard called this new condition 'neurasthenia' to indicate the over-excitation or 'exhaustion' of the nervous system and believed it was particularly common among Americans, saying that American society generated much more excitation of the nervous system than did European society.[15] Indeed, 'American nervousness' or 'Americanitis' became popular synonyms for the disease.[16] Freud concurred with the idea that the gruelling unease that he observed in patients was somehow related to the unremitting stress of urban life but thought there had to be more than external factors causing it. He named this condition 'anxiety neurosis' and suspected it was the result of an opposition between an individual's constitution, desires and aspirations and what modern civilization demanded of him or her.

As you probably know, on his quest to find the inner causes of neurosis Freud received and listened to a large number of patients as they lay on a couch in his small practice in Vienna. Freud had been inspired to do this by his friend Josef Breuer, another Viennese physician, who hypnotized his patients and let them talk about themselves during their hypnotic states.

After examining a large number of cases, Freud theorized neurosis as the manifestation of unresolved conflicts that mostly had their origins in childhood and were often connected to traumatic experiences, frequently of a sexual nature. In general, a neurotic was someone who repressed the discharge of some kind of psychic energy that kept trying to emerge. So he continued to listen to his patients to help them unearth those memories, thus letting the reasons for their unresolved distress surface. One of the notable features of this kind of therapy was that a patient's symptoms mostly disappeared when the moment of their first occurrence was evoked and when forgotten unpleasant or traumatic events connected to those symptoms were recalled to memory.

An emblematic example of this mechanism, one that impressed and inspired Freud, was that of Anna O., one of Breuer's patients.[17] Anna had presented with a nervous cough, visual disturbances, paralysis of the left side of her body, as well as some speech problems. Bizarrely, she at some point also manifested an acute form of hydrophobia. For several weeks she had not been able to drink any liquid. Something as innocuous as a glass of water revolted her and made her nervous, but she couldn't explain why. During a hypnotic session, it emerged that once, in the house of an English woman to whom she was paying a visit, she had caught sight of a dog drinking from a glass. The scene disgusted her, but her manners forbade her to say anything to her host. After recalling this episode, however, she was able to drink again.

It might be worthwhile to summarize what subsequently became of the concept of neurosis.

During the course of the last century, as the number of mental ills afflicting the population increased, doctors thought it necessary to list them all in one book. To that end, in 1952 the American Psychiatric Association published a volume called *Diagnostic and Statistical Manual of Mental Disorders* (or DSM for short). Created to help psychiatrists agree on how to define and recognize mental pathology, the book was supposed to work as an instruction manual, listing the symptoms to observe and by which to identify each disorder among a variety of patients. The book, which is now considered the essential reference for everyone working in mental health and involved in the diagnosis and treatment of psychiatric disorders, had a basic aim of unifying the language of diagnosis. Thus by consulting the pages of the DSM, two psychiatrists living in two different cities, or even two different countries, could use the same parameters of diagnosis for patients showing similar symptoms.

'Neurosis' was listed in the first edition of the DSM. In that edition, neuroses were a broad category in which emotional distress manifested itself through various physiological and mental disturbances. In a way, being neurotic was a slight alteration of normal behaviour. The same broad class of 'anxiety neurosis' was maintained in the second edition of the manual published in 1968, but was dramatically dismantled in the third. The shift to the third edition of the DSM marked an important chapter in the history of psychiatry and laid the foundations for the current system for categorizing anxiety and all other classes of mental illness. In essence, the new edition got rid of the term neurosis – and of everything else that conserved a psychoanalytic meaning – and mainly separated panic attacks and panic disorders from other forms of anxiety, principally because they responded to different kinds of medication.[18]

The fourth edition kept this main separation and introduced new forms of anxiety, each with its own set of symptoms.[19] The classification included: specific phobia, namely fear of a specific object or situation that is usually out of proportion to the actual danger, for instance an exaggerated fear of spiders; social phobia, or social anxiety disorder, the fear of social situations; agoraphobia, the fear of public spaces; post-traumatic stress disorder (PTSD), the manifestation of anxiety in the wake of past exposure to a traumatic event or terrifying threat; panic disorder, in which the sufferer experiences unexpected and frequent episodes of intense fear (like panic attacks); and obsessive–compulsive disorder, characterized by intrusive thoughts and the need to relentlessly pursue a thought or action to get rid of a fear – for instance, having to wash your hands obsessively because you are afraid of catching bacteria.

Another category is generalized anxiety disorder, or GAD for short. Were you to read through the diagnostic criteria for GAD you might conclude that they apply to everybody you know including yourself. Indeed, the DSM says that in order to qualify for a GAD diagnosis, you need to experience the following: excessive and difficult-to-tame anxiety and worry 'occurring more days than not for at least 6 months about a number of events or activities (such as work or school performance)'. You should also manifest three or more of the following symptoms: 'restlessness or feeling keyed-up or on edge; being easily fatigued; difficulty concentrating or mind going blank; irritability; muscle tension; sleep disturbance (difficulty falling or staying asleep, or restless unsatisfying sleep)'. The worry should also not be about something specific and should 'cause clinically significant distress or impairment in social, occupational, or other important areas of functioning' in life.

The DSM is supposed to facilitate the detection of disorders in people who seriously need medical support. But, given these criteria, who wouldn't qualify for such a diagnosis? In a way, GAD is therefore closest to the condition formerly labelled neurosis and represents the ordinary type of anxiety that creeps up on us on a regular basis.

Of note is that, clearly, GAD and the other types of anxiety listed in the DSM are all arbitrary constructs of psychiatrists, illnesses generated by the medical establishment and based on clinical symptoms, not their biology. The disorders are monolithic entities for convenient diagnosis that by themselves tell us nothing about the individual's experience of the disorder.

It is important to remark that at both the symptomatic and the biological level there is considerable overlap across the diagnoses. The various forms of anxiety share their primary neural substrates. Similarly, genes underlying the manifestation of one form of anxiety also play a role in the manifestation of another form (I will talk about this again in more detail in the next chapter).

## Fear conditioning

Unfortunately, a bout of anxiety doesn't always knock at our door. It ambushes us when we least expect it. However, it usually needs something to set it off and such a trigger can often seem innocuous.

Also, worry begets a cascade of other worries, and hearing about recession, or any other trigger, can revive deeper concerns, which are often connected to memories of traumatic events, or more broadly to other unresolved conflicts or problems in our lives. The mechanism of the association between a trigger and the subsequent arrival of a fearful response has long been at the centre of research on fear and anxiety and is related to general theories of behavioural conditioning, which explore how organisms learn to behave in a certain way as a response to changes in their environment.

You may be familiar with the famous experiment of the drooling dogs, conducted by the Russian scientist Ivan Petrovich Pavlov, who in 1904 was awarded the Nobel Prize. Pavlov was using dogs to study the function and mechanisms of the digestive system. Just as our mouths water when we are in front of a succulent meal, when a dog encounters food its saliva starts to dribble. One day, Pavlov noticed that when he or his colleagues visited the dogs in the laboratory, the dogs started to salivate even when there was no food for them. It turned out that the dogs were reacting to the lab coats. Whenever the dogs were given food, the scientist offering them a meal was wearing a white lab coat, so the dogs had learnt to associate the white coat with the arrival of food. Later Pavlov changed the stimulus and struck a bell each time the dogs were fed. After a while, each time the dogs heard a bell, even in the absence of food, their saliva drooled.

A typical laboratory fear-conditioning experiment goes like this: a rat or a mouse is placed in a cage and exposed to a trigger, often a buzzing tone, after which the animal receives a mild electric shock to the feet. The buzz works to *condition* the rodent to the arrival of the next shock. After a few of these pairings, the buzz acquires aversive properties and when presented to the animal it brings about typical behavioural and physiological fear responses. Most often, as soon as it hears the sound, the scared animal anticipates the shock by freezing.

A rodent's fear responses are similar to those of humans. We, too, freeze in our tracks. Imagine your reaction, for instance, when you hear your boss or partner pronounce those four laconic words: 'we need to talk'. If you are anything like me, the normal reaction is to freeze for a moment like those caged rats, because we can be quite sure we are in for some trouble. Then blood circulation races, the heart starts to pound and so on, as described above. Our attention and concentration focus, we are on alert. For many of us, this is because the last time we heard those words we probably had a memorable fight. Those four words function like the buzz in the fear-conditioning experiment. Particularly if reminiscent of traumatic events, external cues like those threatening words can function as conditioned stimuli that trigger a variety of anxious responses. All of this takes energy. Fear and anxiety are draining.

## Anxiety in the brain

Although conceptually distinct, fear and anxiety share their anatomical position in the brain, and twenty years and more of research have mapped their underlying neural circuits, almost down to the single neuron.

The main region involved is the amygdala, the name being a Greek word which means almond. Appropriately shaped, the amygdala is located at the base of the brain, in the temporal lobe (Fig. 8). To have a better idea of where the amygdala is, imagine an arrow that goes straight through your eye and another that goes through your ear: their point of intersection is the position of the amygdala. The amygdala lies at the core of our emotional life, especially our fearful reactions. If we didn't have it, we would probably not be scared of anything! Similarly, an impairment in amygdala function prevents us from perceiving emotion. Indeed,

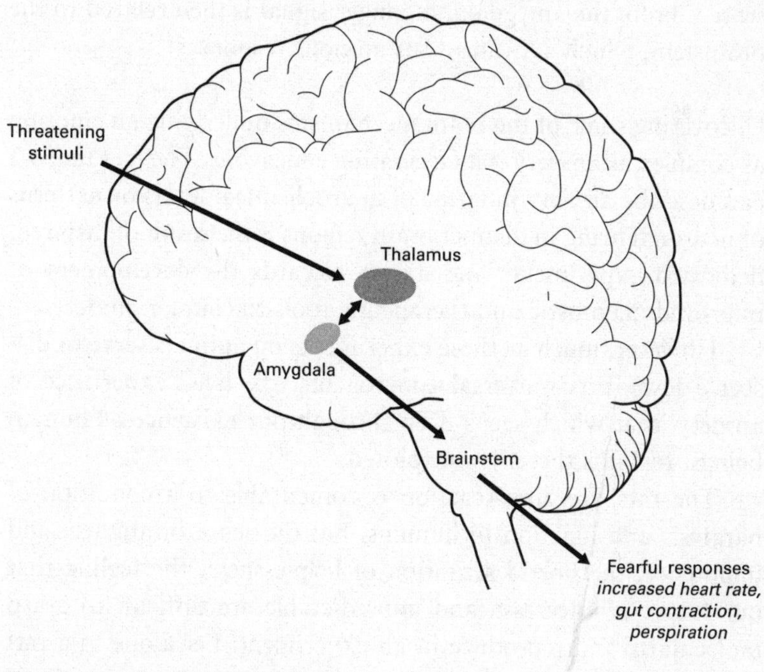


Fig. 8 Anatomy of fear and anxiety

a patient with rare lesions in both her amygdalas (we have one in each of our two brain hemispheres) just could not recognize fearful expressions in others' faces.[20] Despite its small size (about that of your thumbnail), the amygdala has an intricate structure and consists of different parts, each with a different function. For now, just bear in mind that it has a core, called the central nucleus (CeA), and a more external part called the basolateral complex. A conditioned stimulus from the external environment – like the buzzing tone in the fear-conditioning paradigm – first reaches the thalamus, the part of the brain which serves as an integration centre between the outside world and our perception of it. From the thalamus, it travels to the audio-visual cortex where it is processed. But the signal can also follow a shortcut that leads directly to our emotional centres. Indeed, the thalamus has a direct connection to the amygdala, and to be exact, to the basolateral complex. It is here in the amygdala that the emotional memory of the buzzing tone, or whatever our own emotional trigger may be, is stored. From the amygdala, a danger signal is then relayed to the brainstem, which activates your anxious responses.

Uncovering some of the brain mechanisms underlying an emotion as complex as anxiety is a fascinating endeavour. The fact that we can describe anxiety in terms of neurochemical levels or patterns of neuronal firing in distinct brain regions is the result of inspired, dedicated experiments and a step towards the development of improved diagnostic and therapeutic tools to counter anxiety.

However, much as these experiments on animals serve to dissect a few of its universal components, the lived experience of anxiety, that which seeps deep through our existence as human beings, remains thereby unexplored.

The rats' freezing reaction is comparable to a condition of paralysis and inaction in humans, but the sense of anguish and impotence, the horrid sensation of helplessness, the feeling that our future is uncertain and unpredictable are difficult to grasp molecularly and reproduce in an experiment. Let alone in a rat! Ultimately, anxiety is also the manifestation of a tacit awareness that something is missing or wrong in our lives, or that our values and aspirations are out of focus or under threat.

Such contrast between scientific investigation and experience is central to the study of emotions. Science provides an outer picture of the scaffolding of emotions, constructed from universal, measurable and reproducible facts, whereas our direct experience of emotions is much akin to living inside the building behind the scaffolding. It is the fruit of our consciousness, or what is otherwise known as phenomenology, and is not entirely amenable to the scrutiny of science.

## The wind of anxiety

Knowing the limits of science in exploring anxiety as an internal human condition, I was still in search of ideas and experiences I could identify with, in order to make sense of it. Eventually, I turned to philosophy, and particularly to the field of existential philosophy. This branch of philosophy is concerned with how we, as human beings, act, feel and live in search of meaning for our existence. For existential philosophers, there is no rigid, unconditional theory that defines us. Existence prevails over any kind of essence. Indeed, existentialists reject the primacy of universal laws, such as those of science, believing that we are born to seek and choose purpose in a world which is often messy and disorienting. Similarly, we constantly need to find our own values and our own meaning for our lives.

Of all existential thinkers, German philosopher Martin Heidegger (1889–1976) is the one who had the biggest impact on me. Heidegger is best known for having written *Being and Time* (1927),[21] which is considered one of the most influential works of philosophy of the twentieth century. The relevance of Heidegger's thought to the understanding of emotions becomes apparent if we consider the distinction he makes between two main ways of looking at the world, for which he adopted two interesting and innovative terms: *Vorhandenheit* and *Zuhandenheit*. *Vorhandenheit*, which roughly translates as present-at-hand, is a theoretical understanding of reality. It is how we observe and theorize about things, and how we come to know facts about the world through disinterested examination – the way a scientist would. *Zuhandenheit*, or ready-to-hand, is about how we engage with the world – how we are connected to it through our interactions with objects and people in various circumstances. Heidegger accorded the latter greater power, which is to say that our experience of the world overshadows our scientific knowledge of it. It is what comes first, how we initially get to know the world. Likewise, one could say, our experience of our emotional life prevails over our theoretical grasp of it. Heidegger believed that science cannot fully grasp the lived experience of anxiety.

The idea that anxiety and fear are distinct was clear to him. As he wrote, fear and anxiety are 'kindred phenomena' that are often mixed up, but need to be distinguished. Something threatening is 'fearsome' if it is encountered as a definite and real entity. By contrast, 'that in the face of which one is anxious is completely indefinite'. Anxiety does not know what it is anxious about, because the threat is 'nowhere' in particular and has no identifiable source.[22]

Heidegger granted anxiety high importance. In much the same way that we need to be able to experience fear in the presence of real danger in order to survive, for Heidegger we need anxiety in order to 'exist at all' in the world. How is that? Daily we navigate in the world enmeshed in its net of things, people, actions and circumstances. We get up, take our kids to school, go to work, meet our colleagues and friends, go to the gym or the pub, plan a holiday, buy a new piece of furniture for our home, a new CD or the latest phone, and play with our iPad. We are completely absorbed by all this. Heidegger calls this absorption into the world 'falling'. In simple terms, we 'fall' into our routines and, so doing, we tend to overlook, and to stop searching for, the authentic meaning of our lives. Lodged in the 'inertia of falling', we turn away from ourselves. We flee from a meaningful life, because it's easier to do so. We repress anxiety, but 'anxiety is there. It is only sleeping.'

When it awakes, though, our symbiotic rapport with the world fades. In anxiety, those same things, circumstances and people in the world become irrelevant and disappear. Everything 'sinks away'. Any previous connection with the world, and any interpretation of it, is put into doubt. It is no wonder that to convey the disquieting feeling of anxiety, Heidegger also used the word *unheimlich*, which means to be out of home, or 'estranged' from home.[23] In a bout of anxiety, we are forced to become more self-aware and, in so doing, we reconsider the importance of some of the things that we used to hold so dear and our engagement with them. We question ourselves. Anxiety discloses the world and our condition in it as they are, void of superfluous adornments.

Our anxiety also connects to the future. We are human beings who exist in time, Heidegger insisted. Indeed, we are not anxious about what has happened, or what is about to happen. Instead we become anxious primarily about what *may* happen. Worry often creeps in when we think about the endless chances that we may or may not seize in life. Anxiety is rooted in the realization of our freedom to choose who we want to be and how we want to live. For Heidegger, choice comes with a substantial difficulty, because it is, profoundly, about the type of life that makes us more authentic. It is not simply about what job to take, which house to buy or with whom to share a life. It is about the job, the house and the individual that bring out the highest potentiality for our being, upon which we rely for the achievement of our happiness, one could say. There is no fixed recipe. Only we can understand what is best for us. It is about choosing something for its meaning for us alone and not because it conforms to society's norms or anyone else's values.

How many times have we had to face important decisions and been baffled by the possibilities? Sometimes the decision is relatively simple and there are only a couple of options from which to pick. On other occasions there is a lot at stake and the possibilities are less clear. So, for example, think about the time when, on reaching adulthood, you had to settle on a career to follow.

If you were lucky, you may have had a single passion ever since you were a child, a passion you were always able to cultivate. For some, choosing what to be and becoming it involves a more tortuous path. Understanding your true inclinations and following them may be a stressful process.

In truth, being authentic to oneself is an endless challenge that we face, with varying degrees of awareness, day after day. Anxiety is always around the corner; we are in constant negotiation with it.

So, anxiety is simultaneously the starting point on our journey to become our true selves and the awareness that we are alone in an ocean of life possibilities. Dreadful, isn't it?

When I approached Heidegger, I realized that his description of anxiety paralleled my own personal metaphor of anxiety as a robust wind. The wind that had swept everything away, dislodged me from the carousel of life and left me on an empty, shadowy stage, with just one spotlight, directed at me. His words spoke to me as no experiment could. Ideas and philosophy matched my personal feeling of anxiety more closely than did the laboratory and science.

## Take the alternative route

Heidegger definitely made me see the neuroscience of fear and anxiety in a new light, and the consequence was that I began seeking out studies that somehow supported anxiety's useful role in life and offered practical clues as to how to manage it.

Going back to those rats, I found an interesting series of experiments that refined the original ones, taking them a step further. In one, conducted by the neuroscientist Joseph LeDoux and his colleagues at New York University, rats conditioned to the buzzing tone were given the opportunity to move to another room while the tone was being emitted. If they chose to enter the new room, the tone stopped and the shock did not ensue. After a few such repetitions, the animals learnt the advantage of their new behaviour – choosing to change rooms – and that discovery in turn altered their fear responses. The danger signal stored in the amygdala did not reach the brainstem and did not trigger the freezing reaction. It went instead to motor circuits and incited the rat to take novel actions.[24] What is indeed remarkable in this set of experiments is that the flow of information is effectively rerouted only if the rats take action, and not if they remain passive. It is clear that there are two distinct neural outputs from the amygdala that mediate the impact of the tone, one triggering passive fear reactions, the other facilitating novel actions (Fig. 9). In rodents and humans alike, both pathways are available, but the second one has to be learnt. By engaging this alternative pathway, passive fear is replaced with action, what in the field is known as an active coping strategy.

In 2010, two colleagues from the laboratory of Cornelius Gross at the European Molecular Biology Laboratory where I used to work deepened these findings, in conjunction with other collaborators. Using a combination of genetic technology, fMRI imaging and behavioural tests, they were able to map the specific neurons in the amygdala that are involved in the neural switch

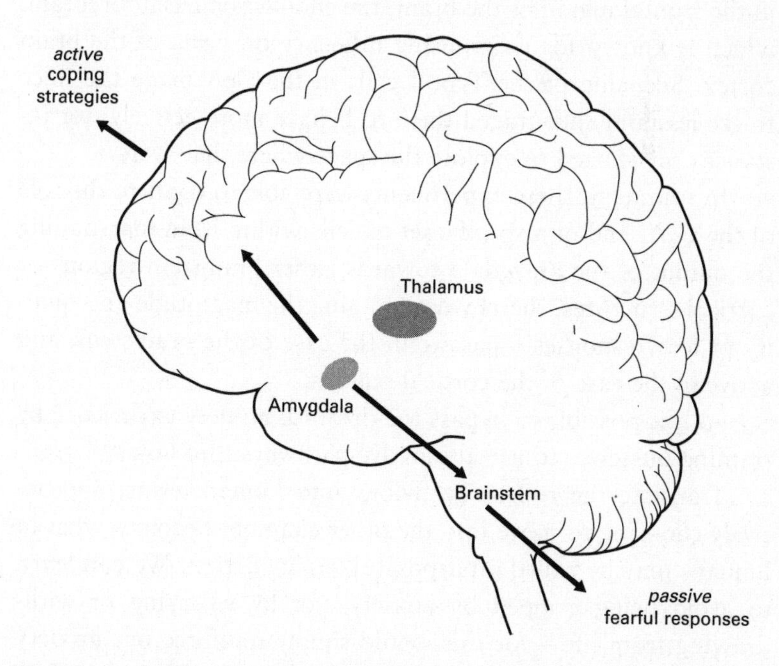


Fig. 9 Reversal of neural pathways facilitating active over passive fear responses

from passive to active fear.[25] To do this, they created a transgenic mouse. This is basically a mouse designed to have high amounts of a particular protein in a specific brain region of choice. The region of choice in this case was the central nucleus of the amygdala (CeA), because they wanted to further explore its role. The protein in question was the serotonin 1A receptor (Htr1a). Sitting on the outside of neurons, receptors are molecules that are targets for neurotransmitters. The Htr1a is special because it has inhibitory transmission activity, which means that if a molecule binds to and activates it, neural activity is suppressed, hence less anxiety.

My colleagues gave the mice a drug to selectively silence the CeA. What they observed is that only a specific subset of cells in the CeA responded to the drug. These were called Type I cells. Mice given the drug were also placed in a magnetic resonance scanner to see what happened following inhibition of neuronal activity in the CeA Type I cells. The researchers discovered that inhibition of activity in this type of neuron was linked to activity in the frontal region of the brain, the cholinergic basal forebrain, which is known for its arousing influence on parts of the brain cortex. Silencing of the Type I cells in the CeA made the mice freeze less and encouraged them to behave more actively: for instance they started to explore the space where they were.

In summary, these experiments were able to confirm the role of the CeA, and of a specific set of cells within it, in marshalling the output of the amygdala towards either brainstem regions or cortical structures, thereby determining the magnitude and quality of fear responses – passive in the case of the brainstem, and active in the case of the cortical regions.

So it is possible to bypass the dreadful anxiety experience by training ourselves to use alternative pathways. But how?

Translate the rodent behaviour into human terms, and actively choosing to move into the other chamber becomes what in humans may be called a purposeful coping action. We can learn to avoid being gripped by anxiety, not by worrying or withdrawing from life – for this would simply reinforce our anxiety symptoms – but by actively turning away from negative thoughts, engaging in pleasurable activities and adopting constructive behaviour.[26] Precisely what you do does not matter so much as the fact that you do something which distances you from your concerns and that you concentrate on something positive. So, listen to your favourite music, take a stroll or write a letter to a friend, meditate. We all have different pleasures. None of this means that you should simply avoid your problems, but that you should reach a state of mind that can help you face them with greater awareness.

Such a positive attitude may sound intuitive and straightforward, but is easily overlooked. At times, I have purposefully tried to use knowledge of the brain routes of my fears in the attempt to reinforce it, almost by translating the mental imagery of those neural crossroads into resolute choices: I won't let fear take its usual course, I'll divert it. I can't say it works better than just telling myself to calm down – or reminding myself of Heidegger's thoughts, for that matter – but it contributes to achieving a positive mindset.[27]

The night the world markets slumped, going out for a drink with Robert was certainly better for each of us than staying alone in the house at the mercy of convoluted, senseless fears. Talking and sharing concerns with someone else halves the weight of those concerns and can be inspiring. Together, we were able to filter out some of our irrelevant negative thoughts. Remarkably, Heidegger's insights find a few parallels with what we are learning from psychology and neuroscience.[28] In their different ways, both visions could be taken as incitements to actively engage in positive endeavours that can help us both in the short- and the long-term management of life.

As emphasized earlier, although seemingly innocuous stimuli can trigger anxiety, on a deeper level anxiety is the result of losing focus on the personal values and life choices that form the core of our existence. So, hearing about the crash of the stock exchange is really only a cue, a spark, for the ignition of deeper conflicts, and the bewildering set of worried reactions that can result is a message from our body that we need to solve them. Paradoxically, we worry because we think that worrying is the only helpful strategy, and indeed that worrying keeps us safe (because it keeps us from taking action). In truth, it merely keeps us preoccupied, without getting us anywhere. At first, reacting to anxiety might feel like a titanic enterprise. But in time your brain can learn how to shift your attention away from the worry.

## The plastic brain

I know all this sounds more easily said than done. In some cases, anxiety can be seriously paralysing and relentless. Those who struggle with its numbing effects may take a long time to learn how to free themselves of them. Not for one minute do I want to diminish the magnitude of such problems or the distress they cause. Take, for instance, individuals who have experienced trauma, whose fear-eliciting memories control their mental life and behaviour.

Nearly a year after the London tube bombings in July 2005, Thomas, a computer scientist who had survived the explosion in the tube carriage at Edgware Road, couldn't sleep properly at night.[29] He kept having nightmares about the explosion, and memories of that terrible morning still haunted him. Moreover, something unusual and specific kept occurring. Whenever he laughed, he immediately became sad. His laughter at something funny would be followed by low spirits.

A therapist in the NHS helped him to revive his experience of that day and retrieve the memory of the moments preceding the blast. When Thomas boarded the Circle line train he sat and opened a book by Vladimir Kaminer, one of his favourite authors, who writes funny stories. Thomas recalls that as he sat, only a few feet away from the suicide bomber, he was in a good, joyful mood. Instants before the terrorist blew himself up, Thomas might well have been laughing. So what happened is that his mind registered that sequence of events and replayed it in the months subsequent to the attacks. Hence, laughter easily turned into despair.

The atrocious events of that summer morning of 7 July 2005, the day after London won the bid to host the 2012 Olympic games, irrevocably affected the lives of those involved. Fifty-six people, including the suicide bombers, lost their lives. Hundreds of people were physically injured, some losing limbs or becoming permanently paralysed. But, as in the case of Thomas, the events also inflicted invisible injuries in the memory of many of the commuters who travelled on those trains. The powerful recollection of the experience of deep fear and feelings of helplessness and horror in the face of a life threat is the core symptom of post-traumatic stress disorder (PTSD). One of the hallmarks of PTSD is a persistent sense of threat that endures despite the danger having passed. This constant condition of hyper-vigilance and exaggerated emotional reaction can certainly get in the way of a normal, functioning professional or social life. For people with PTSD, the alarm of anxiety never fades out. They are continuously expecting the danger to emerge from around the corner.

However, remarkably, in the face of incredibly disquieting events – terror attacks, war, or natural disasters such as earthquakes – most people react differently. For the majority – especially those who have been only indirectly linked to a traumatic event – the stress symptoms are temporary, decline over time and don't have lasting mental health complications. Basically, people tend to carry on with their lives.

Why is it that a traumatic event can leave a deep, indelible mark on a few individuals and almost no trace in others? The answer lies in a multitude of factors.

Some are rooted in our past and personal life stories. Our tendency to express anxiety in adulthood is the outcome of mechanisms developed in critical formative portions of our lives. Such mechanisms depend strongly on the type of environment we experienced. But there are, of course, also biological factors that, in combination with the environment, differentiate our dispositions and reactions to the external world. For instance, some amygdalas are more readily active and excitable than others, which makes their owners more sensitive in the processing of emotions and more sensitive in their responses to given circumstances. Genetic variation certainly plays a role. For instance, it has been shown that people with particular subtle differences in the gene sequence of the Htr1a have decreased amygdala reactivity, which confirms the role of this receptor in fine-tuning anxiety responses.[30] The cumulative effect of past lived experience and biological disposition makes people either more or less resilient to adversity.

Everyone who was in London on 7 July 2005 will recall their experience that day and will have come to terms with the events in their own way.

That July morning, right at the moment the bombs were detonated, I was already sitting at my desk in the office. Only half an hour before, I had jogged past the spot where the number 30 bus blew up, on my way towards Russell Square. My most vivid memory of the day was the quietness that sank throughout the city in the evening. I had never seen London so silent and so sad. For several weeks after the events took place, I was hesitant to board any train or bus in London. Luckily, I didn't have to take public transport to go to work every morning because my daily commute to the office consisted of a short walk from Fitzrovia to Covent Garden. In general, whenever I could, I made every journey on foot. I also avoided crowded spaces, thinking they could be targets of new attacks. Indeed, in the year immediately following the bombs there was a reduction of about 15 per cent in the overall use of public transport in London.[31]

At the end of summer, my levels of anxious anticipation regarding the threat of new terror attacks lowered and I started to use the tube again. But I continued to be vigilant. I confess that, once or twice, when I saw someone carrying a backpack, I left the carriage. I could not help it. Over time I taught myself to assess the risk of a terrorist attack and acknowledge that it made no sense to let any anxiety about an attack get in the way of my leading a normal life.

In his seminal essay 'What is an emotion?', which I mentioned earlier, William James made a clear statement of how to exercise control of our emotions: 'If we wish to conquer undesirable emotional tendencies in ourselves, we must assiduously, and in the first instance cold-bloodedly, go through the outward motions of those contrary dispositions we prefer to cultivate.' If we have a tendency to cultivate anxiety, we need to diligently exercise our capacity to counteract it with calm and positivity and we must start from the body. 'Whistling to keep up courage is no mere figure of speech,' James says.

The brain and its neurons are incredibly plastic and each purposeful action towards change, however small, contributes to the consolidation of new behavioural patterns and of the underlying neural circuits that bypass your anxious reactions.[32]

We can condition ourselves to correct our behaviour and avoid falling prey to anxiety. Little by little, the non-fearful strategy will establish itself as the preferred neural pathway in our brain and, as a result, we will be better equipped when anxiety approaches. Before anxiety takes a complete hold of us, we will be able to 'switch on' the alternative pathway. Making use of brain plasticity is like following a different route to reach your destination. Imagine that you have always taken the same path to reach a lake in the middle of the woods. One day you notice that not far from your customary path, hidden in the bushes, there is an unbeaten track that someone has started to carve out and you decide to follow it. At first, the new track is full of bumps. But the more you walk it, the more it will widen and flatten. With time, the new track will turn into a road and you will make it your preferred route.

Our growing knowledge of the phenomenon of neuroplasticity has reframed too the way we understand psychoanalysis and other forms of psychotherapy. It is now clear that 'talking cures' of all kinds are not mere intellectual exchange, but a biological treatment that directly affects our brain. Several brain-imaging studies in which scans were made before and after therapy have shown that the brain indeed rearranges itself during the treatment. Strikingly, the more successful the therapy, the more profound the changes in the brain. The recall of memories, their elaboration and the conscious refocusing of attention on new behavioural patterns produces durable biological changes in the brain: synaptic connections grow and modify, new neuronal connections are made. A new mental reality is established. For instance, an fMRI brain scan revealed that four weeks of therapy normalized the hyperactivation of the amygdala in patients experiencing panic disorder.[33]

Cognitive behavioural therapy is based on the assumption that anxiety is caused by cognitive distortions, that is, unrealistic or exaggerated thoughts – like worrying that a gunman might enter our regular café and shoot all the customers on a peaceful Sunday afternoon, when we know that although such an event is possible, its likelihood is truly slim; or, as in the case that sparked this chapter, fretting that not having money invested in the stock exchange might put you at risk during a financial crisis.

One piece of advice from cognitive behavioural therapy is to recognize these distorted cognitions, localize our fears, evaluate them from a distance and identify all the reasons why some do not make sense, ungrounded as they are in reality. In other words, it teaches us to give fear the reasons, if any, that it is looking for. If the thought of the gunman prevents us from going to our regular café, a therapist will encourage us to overcome that irrational fear and to observe and experience the safety of the action of entering the café. By asking us to replace old and negative behaviours with new ones, and to practise novel actions, therapists are encouraging us to use the plasticity of our brains. With the help of a skilful and knowledgeable consultant, psychotherapy digs deep into our brain like a neurosurgeon. We leave the therapy room renewed. Not only because of acquired awareness about our past and present behaviour patterns, but also because that awareness is underlined by ongoing chemical transformation in our brain.

## Trips into tranquillity

In summer 1958, visitors at the convention of the American Medical Association in San Francisco were confronted by the gaze of a 20-metre-long wormlike creature made of parachute silk, that regularly 'breathed' as to mimic the movement of a caterpillar. This was an installation created by Salvador Dali, that was designed to be viewed on the inside as well, where there were four human figures. On first walking into the caterpillar's interior, visitors saw an emaciated man who held a staff capped with a black butterfly. For Dali, this depicted human anxiety. Next was an almost transparent woman who also carried a staff topped with a moth. The third figure was a maiden with a head full of flowers, whom Dali called the 'true butterfly of tranquillity'. Finally came another maiden who skipped rope towards serenity.

The caterpillar creature, which was not calming at all, was an artwork commissioned by the Wallace Laboratories, the manufacturers of Miltown, a drug discovered by accident in 1955 (first intended as a muscle relaxant) that was to become one of the most consumed minor tranquillizers in the early history of psychopharmacology.

The eccentric surrealist artist, well versed in portraying the mind and particularly the subconscious, named his creation *Crisalida*: 'The outer structure of Miltown is that of a chrysalis, maximum symbol of the vital nirvana which paves the way for the dazzling dawn of the butterfly, in its turn the symbol of the human soul.'[34] It seems Dali was sufficiently familiar with the experience of human anxiety to visualize it with confidence and represent the drug as a journey through the interior of a seemingly disturbing and terrifying creature. The journey began in anxiety and ended with the arrival at a place of harmony. According to the narrative of *Crisalida*, Miltown paved the way to an ataraxic state of mind distant from the turmoil of anxiety. When *Crisalida* was shown, profits from Miltown had already reached incredibly high peaks. Those were *anni mirabiles* for drug companies, which found fertile ground in a society needing to cope with the stresses of modern life.[35]

Later on, another class of anti-anxiety drugs, the benzodiazepines, were introduced into the market. These achieve their calming effects by binding to the brain's GABA receptors, which, like the Htr1A receptor mentioned above, have inhibitory transmission activity. They slow your racing heart and rapid breathing almost instantly and they act fast to pacify your anxious thoughts. They are effective and relatively cheap to produce. These drugs were marketed to target a specific portion of the population. Some of the promotional material directed at doctors portrays ordinary individuals who seem to be in need of the advertised drug to overcome worries arising from a variety of everyday hurdles and difficulties in social or interpersonal contexts.[36] The range of characters depicted encompasses the housewife who cannot cope with daily household chores, the person who cannot make friends and the manager or high-finance banker who, like Robert, is under pressure at work. A frequently recurrent figure is a woman troubled by tension and anxieties arising from difficulties in her household. Males are more often portrayed in their working environment, dealing with a multitude of business-related challenges demanding high standards of performance, multi-tasking skills and a range of social and interactive skills – for clinching important deals and for a successful integration in the office. The press started to label the drugs with catchy terms that gave readers and potential consumers an idea of what to expect of them. They were, for instance, 'Peace of Mind Drugs', 'Aspirin for the Soul', 'Happiness Pills', 'Mental Laxatives' and even 'Turkish Bath in a Tablet'.[37]

However, soon after their commercialization and wide dissemination, the minor tranquillizers had become regarded as both an opportunity and a danger for society. The pharmaceutical industries were accused of practising 'mystification', in presenting problems that form part of normal human existence as conditions requiring medical attention.[38]

Benzodiazepines never disappeared from the market. Nowadays, they are still amongst the most widely prescribed medications. You can meet people who gulp down a Xanax before boarding a plane, or before an important interview. I have never taken Xanax or the like. Whenever anxiety has knocked on my door, like that night when my friend Robert called me, I have resorted to different kinds of remedies. I have tried yoga, camomile tea, alcohol or a chat with a friend.

I don't have any strong objection to the use of medications in principle. A world without synthetic drugs is unthinkable. Anti-anxiety drugs are indeed effective in helping you manage anxiety in the short term. Perhaps I have never used them because I have never felt my anxiety reach a level that I could not manage alone. Or maybe it is because, stoically, I have been under the conviction that I should work through my feelings without intervention, or that any perturbation in my mood should be first faced by trying to find the inner causes for it. Pills are attractive. When everything else fails, they are a ready alternative. And, for sure, anti-anxiety drugs have become easily accessible, the only hurdle being to obtain a doctor's prescription. However, they can generate dependency, too. If you benefit from the use of an anti-anxiety drug in a particular difficult situation, each time that same type of situation arises, you will be tempted to take the drug. It will give you relief on that occasion, but it won't prevent a recurrence of anxiety.

What remains disputable, however, is whether medications should be prescribed to people whose anxiety is not such as to render their lives dysfunctional: just the regular, all-too-familiar anxious edge that afflicts almost everyone and for which boundaries of diagnosis and treatment are extremely murky. This is the type of anxiety that affected Robert, and that regularly affects you and me. It is the type of anxiety that in our current society has been labelled 'generalized anxiety disorder' and is so widespread.

The aspiration to a carefree or tranquil state of being, the idea underlying Dali's *Crisalida*, is perhaps intrinsic to the human condition. We dream of an anxiety-free existence, or at least intervals of careless living. Those who suffer from anxiety and those who work to alleviate its burden, be they laboratory or clinical researchers or drug companies, are all focused on obtaining relief. However, what this attitude does is to load anxiety with negative connotations and depict it as an undesirable condition. Anxiety is portrayed as an unwanted and avoidable psychiatric condition that requires intervention. At some point in *The Age of Anxiety*, the poem which I remembered back in that early-morning bar, Auden describes anxiety as a bad smell that haunts 'the minds of most young men', under the illusion 'that their lack of confidence is a unique and shameful fear which, if confessed, would make them an object of derision to their normal contemporaries'.[39]

Anxiety is still flagged as something to be avoided, even to be ashamed of.

A fundamental problem with psychiatric categories is their dependence on social context. Whatever their nature and whatever name they are given, you cannot divorce them from the context in which they arise.

The French historian of science Georges Canguilhem (1904–95) formulated a general framework by which to explain this constant tension between the disorder itself and the meaning attributed to it, which also helped him come up with a distinction between normality and pathology. He believed that every organism, and therefore every individual, comes with its own set of properties and functions – its own general physiology, if you like – which allows it to function in the world and adapt to the environment. This for him constitutes the intrinsic vital normativity of an individual. From this perspective, even impairment in the regularity of the body constitutes the normativity belonging to that individual, as part of his or her vital norms. But outside the organism, in society, exist other norms that deem certain kinds of behaviour unacceptable or problematic. This constant tension between vital norms on one side and social norms on the other underlies the emergence of new psychiatric conditions, the labelling of normative aspects of our physiology as lamentable disorders.[40]

Contemporary societies, especially in the Western world, with their reverence for values such as self-sufficiency, initiative, achievement, relentlessness and efficiency, have become less tolerant of anxious states, both disruptive and mild, and have transformed our expectations of individuals. There is neither time nor patience for anxiety. These values act as a norm against which differences or deviations, such as lack of energy, low mood or resignation, and individuals displaying them, are judged as pathological. Hence the growing efforts to eradicate anxiety, as reflected in the relentless rise in prescriptions of anti-anxiety medications and the search for new and even more effective drugs.[41] So, ironically, we live in a society that generates anxiety but overlooks its existentially positive role. Moreover, it is our efforts to rid ourselves of anxiety that may propagate a culture that keeps on producing it.[42]

## Coda

Robert eventually did lose his job as a banker. It wasn't at all easy for him at first. During his period of unemployment I met him several times to try to help him figure out the next steps. In our full-ranging conversations about science, philosophy and life in general, we agreed that whilst we are powerless to affect the larger economy we can certainly change our reactions to it.

Occasionally, we found ourselves pondering what life would be like if these current years in our young adulthood were not affected by the crisis. With resignation, we wondered whether we were perhaps born too late. But then, even if we bow to the fact that we may live in an age of greater and disproportionate anxiety, there is truly not much we can do. These are the times we live in and we need to make the best of it.

Except in the case of a blatantly serious traumatic experience leaving a lasting mark in our memories, most of our anxieties actually reside in our constant desire to change our identities and in the realization of the impossibility of finding any *definite* guidance for our actions. Anxiety is being unable to cope with uncertainty, and in severe cases the experience can be terrifying.

But isn't life nothing but uncertainties?

Despite the world's attempt to stigmatize anxiety, we should cherish it. We need anxiety in order to make an objective assessment of our existence and, thereafter, to make a significant change, strive for something positive. Anxiety is our amber light: an opportunity to make the right choices and to identify the goals and actions that we consider worth pursuing if we are to live authentic lives, or at least lives that have meaning for us. These goals will, of course, vary. But whether you long for a large family, a job in sales, a career as a musician or the perfect body, you need to be determined. So, when the howling wind of anxiety blows, we need to be robust enough to remain rooted in our aspirations, but flexible enough to attune to the types of change we need. Anxiety is an opportunity to make the indefinite into something more certain, vagueness and dimness into precision and clarity. If you can do that, then anxiety will fade and more positive emotions will set in.

To my joy, Robert bravely used part of his retirement savings to start a new business, a café and bookshop in the heart of Soho. Having always dreamed of owning such a shop, he seized the opportunity the recession gave him to nurture his passion for books. Now, whenever he or I need to exchange views on life, I go and visit him there.

Several among those who lost their competitive jobs in high finance during the recession came up with creative ideas for an alternative occupation, rediscovering some of their old, long-repressed passions and dedicating themselves to them. Who knows if Robert's café will sustain him through the economic crisis. Yet his triumph is to have accepted that life is all about uncertainties. Fear and bravery are two sides of the same coin. Bravery is to go ahead with your actions despite the fear, and to turn corners without knowing what lies ahead.

When I need to remind myself of this, I evoke a beautiful text written by the Austrian-Bohemian poet Rainer Maria Rilke. It talks about the 'fear of the inexplicable', a phrase which in itself encapsulates the meaning of anxiety. He says:

> . . . fear of the inexplicable has not alone impoverished the existence of the individual; [ . . . ] it is shyness before any sort of new, unforeseeable experience with which one does not think oneself able to cope [ . . . ] For if we think of this existence of the individual as a larger or smaller room, it appears evident that most people learn to know only a corner of their room, a place by the window, a strip of floor on which they walk up and down. Thus they have a certain security. And yet that dangerous insecurity is so much more human . . . [43]

Rilke is encouraging us to embrace uncertainties. Accepting and learning to cope with that fundamental and intrinsic aspect of existence is the best approach to living with and through anxiety.

peel the crop, eating one or two of those delicious fruits, while Nonna takes a moment to look out to the sea. A strip of water is visible from her kitchen window. Granddad once bought her a telescope and placed it pointing at the sea so that she could spot him as he passed by in his boat during his weekend fishing trips. Always around noon. All Grandma had to do to see him was to peer into it to catch the boat coming into sight.

After all these years I still don't know if the telescope was there to reassure her, or if Granddad secretly didn't mind having a private coastguard, checking that everything was all right. In either case, his passage was also a signal that he was on his way back home and about to reach the pier, and that it was time to crush the herbs, boil some water and, when the grandchildren were around, to summon us to lay the table and go and pick fresh parsley, lemons and sage from the garden.

The telescope is still there, but the boat no longer comes into sight, except in the crosshairs of her memory, the memory of a life spent together that lasted for sixty years. Now, there is no more freshly caught fish to bone, neither is there a packed breakfast to prepare for Granddad's dawn fishing expeditions. Only a proud portrait of him at the helm of the boat hanging on the wall. Underneath it, fresh flowers and a candle. Nonno Nino died in June 2007 at the age of eighty. He lost his life to stomach cancer, after a battle that lasted a little over a year and involved two surgical operations, a lot of bargaining with hope, and a lot of courage. The entire family was saddened by his departure, but for Lucia, his spouse, the separation was tougher because she lost her life companion. When my grandmother stares at the sea from the window, she reaches a place known only to her. 'Grief makes an hour ten,' said Shakespeare. 'Suffering is one very long moment,' echoed Oscar Wilde in his *De Profundis*. For those in grief, time proceeds at a different pace, the seasons, days, hours and minutes lagging as if the earth itself had slowed in its rotations. Loss tilts the plane of our existence. It is a disorienting experience, an emotional earthquake, capable of upsetting our compass points of reference as we navigate through what remains of life in the absence of that loved one.

When we grieve, we relive memories of moments shared with the person who has died. At first, memories, even the most joyful, are intrusive and can be extremely painful. The ancient playwright Aeschylus once said: 'There is no pain so great as the memory of joy in grief.' Memories prompt yearning, and a desire for reunion that cannot be met. At best we try to avoid circumstances, places or activities that remind us of the lost one. Over time, however, acceptance of the reality contributes to coping better and memories are what most preciously helps us bring the lost ones close. Proverbially, time is a cure for grief and sadness.

Grief is an intense emotion which can also be regarded as a process, a trajectory that involves other emotions, like knots to untangle in a chain. Grief ages. It is first young and insistent, then calmer and more discreet. Although there is no prescribed or typical reaction to loss, a large number of bereaved individuals share the experience of a few common stages.[2] First comes denial. You just can't believe, nor accept, what has happened to you and that the life of someone you love has been snatched away. The loss is unbearably traumatic and denial works like a convenient filter that only lets in what you are able to handle. Then ensues anger, at yourself or others, for not doing enough, for not being able to prevent the death. Turned inward, the anger at yourself often transforms itself into guilt. Finally, we learn how to live with the loss, we frame it, putting into a more distant perspective, we learn how to deal with memories, reaching a level of acceptance. But before that point, which may take a long time to achieve, there is perhaps the slowest, most painful and fragile stage to go through. That is deep sadness.

## Down in the mouth

The title of this section is a common idiom, in use in the English language since the mid seventeenth century,[3] to label the feeling of being dispirited, dejected, discouraged, disappointed. All these terms express an emotional transformation involving a theft, a subtraction of some sort.

But there is another familiar set of metaphors denoting sadness that involve motion. We sink, we fall, we descend into low spirits or into the doldrums. We feel down, everything loses vigour and declines, as if at the mercy of a faceless gravity. It's a downward movement. An overall contrition, and an inward shrivelling. In his chapter on 'low spirits', Darwin describes sadness as a state characterized by languid blood circulation, pallor and flaccid muscles. 'The head hangs on the contracted chest,' he writes, and 'lips, cheeks and lower jaw all sink downwards from their own weight'.[4]

Indeed, the expression 'down in the mouth' finds its tangible correspondence in the micro-movements of our facial expressions. One of the first perceptible signs of sadness is the drawing down of the corners of the lips, operated by tiny muscles known as the depressores anguli oris. This downward curve of the lips is then accompanied by something else going on in the upper area of your face, which Darwin calls a degree of 'obliquity in the eyebrows'.

A contraction in the orbiculars, the corrugators and in the pyramidals of your nose raises the inner ends of your eyebrows and draws them together, giving rise to a small 'lump'. Even the upper part of your eyelids is raised, assuming a pointed triangular shape. Darwin underlined the power of this particular piece of muscular contraction in the overall delineation of a sad expression. In some, but not all people, the most dramatic effect of such muscle contraction is the formation of deep furrows across the forehead which almost look like a horseshoe.[5] Darwin wrote that these muscles might as well have been named the 'grief-muscles'.

What I find most remarkable about these and all outward physical signs of emotion is their unique, distinctive overall outcome. As we have seen in the preceding chapters, all emotions pervade us with a whole programme of spontaneous bodily changes, facial ones in particular, that come into play as appropriate. It's not easy to fake sadness. The movement of the muscles arching the inner points of your eyebrows is hard to achieve voluntarily, even for actors.

Though skilled at reproducing emotions, even single ones, actors are not infallible at reproducing the singularity of one expression, or all of its components. The man and woman in Fig. 10 are

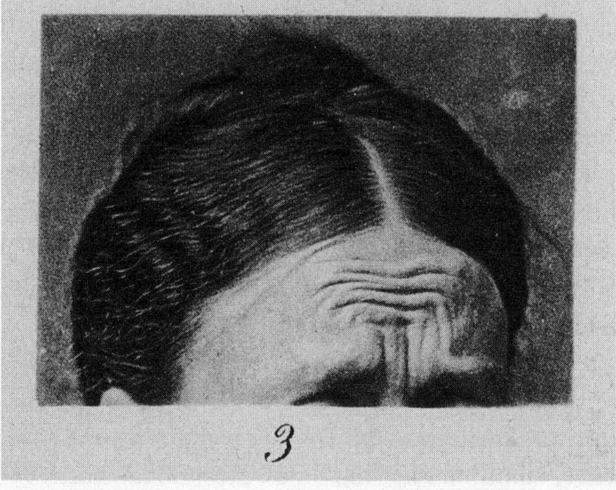


Fig. 10 Pictures of actors simulating grief. Only the man (above) gets the effect on the inner ends of his eyebrows correct. (Wellcome Library, London)

not people in grief, but actors simulating grief, as immortalized by Oscar Rejlander, the Swedish photographer Darwin had hired to illustrate sections of his book. Darwin remarks that in the woman at the bottom of the figure the eyebrows are not engaged exactly as they would be if the sadness were indeed genuine. However, somehow she very successfully acted out the forehead wrinkles. By contrast, the man was better at arching the inner ends of his eyebrows, though not equally on each side.

In sum, it is hard to mimic emotions authentically, especially to an expert eye (I will talk about actors, facial expressions and emotions in more detail in the next chapter). On the other hand, where emotions are genuine they are also difficult to hide.

But something else makes sadness inimitable, and that is the flow of tears.

## Cry me an ocean

Wounds and cuts on the skin are cleaned to avoid infection. Emotional lacerations too need rinsing. Sadness and grief are sentiments awash in emotional weeping. Tears are an excess of sensibility, a loosening balm for our feelings.

A few aspects of the physiology of tears are relatively straightforward. As part of their universal function, tears work simply as an effective saline eye lubricant. If eyes couldn't produce tears, they would be constantly dry and defenceless against external irritants. Lubricant tears are in fact constantly produced and poured on to the surface of our corneas by the lachrymal glands, tiny almond-shaped bulbs located at the inner corner of the eye, close to the nose. To fulfil their protective function, tears also contain lysozyme, which is a natural disinfectant.

But tears are definitely more than salt and an antiseptic when they are shed to bedew our fragile dry ground of sadness – that is, when tears are *emotional.* Despite being a relatively ordinary occurrence, emotional tearing holds exceptional status in evolution as a uniquely human capacity. There is no convincing evidence that any animals cry, even our closest fellow primates the chimpanzees.

Of course it all depends on how you define crying. You can hear juvenile mice, rats or monkeys utter loud, squeaky vocalizations and observe the desperation in their eyes and body movements when they are separated, even for a short time, from their mothers or caregivers. These are clear bold indices of distress. Human babies do that, too. They lament the separation from their mother quite clearly and outspokenly. In all these cases, the crying is the externalization of a protest. But the sad discharge of tears from the eyes in response to loss, or as proof of another kind of emotional shaking, is a feature attributable exclusively to human beings, which even babies learn to do only some months after being born. And this is not just because the lachrymal glands have not developed properly or are not yet functional. Darwin noticed this in his own children. When he accidentally touched with one edge of his coat the eye of one of his children, who was aged just over two months, the child screamed loudly and the touched eye watered, but the other eye stayed dry. Tears started to run properly down the cheeks only when the baby was almost five months old.[6]

What is the purpose of crying? Psychologists have wondered for a long time.

In his *Letters from the Black Sea* the Latin poet Ovid wrote: 'tears at times have all the weight of speech'. Indeed, tears have enormous communicative power. Just as the crying vocalizations shared across various members of the animal kingdom clearly convey a baby's distress at separation from its mother, emotional tears are for humans efficient expression signals of sadness. The psychologist and neuroscientist Robert Provine, who has researched into several behavioural oddities – things like yawning, coughing and hiccupping – put the communicative power of tears to the test. He has demonstrated that tears unequivocally underscore the emotion of sadness. He and his colleagues showed a group of eighty people several pairs of identical portraits of sad facial expressions. For each pair, one portrait had tears and the other had the tears digitally removed. Without exception the portraits with the tears were ranked sadder than their counterparts with the tears erased.[7] In addition, tears worked to cancel out any ambiguity in facial expression recognition. If you remove tears from portraits of sad faces, the same sad emotional expression is more likely to be misinterpreted and is described as disparately as contemplation, puzzlement or a feeling of awe.[8]

Tears are often regarded as a sign of fragility. Indeed, by clouding our vision, tears do make us vulnerable to others. Crying is also an addling experience. Especially if frantic and desperate, crying halts us. It sequesters us into a state of confusion and paralysis from which it is not easy to see or act lucidly. Crying temporarily distorts perception and thereby prevents us from dealing with something for which we have no rational explanation or ready solution. The trade-off for this is that tears can communicate our attachment to and need for others, giving us a chance to strengthen relationships.[9] Vulnerability bonds.

But above all, the release of tears in moving situations is commonly regarded as a cathartic, liberating event. A good cry can get you out of a contrived mood and work as an emotional purifier. The crying episode may be stormy and bewildering, but when things clear up and quietness returns, we all benefit from the shake.[10]

One question remains. What makes emotional tears singular? In other words, are the tears we shed when we smell a raw onion different from those that roll down our cheeks when we say goodbye to someone at the airport? There is no conclusive answer on the difference in chemical composition between the two types of tears. Provine has speculated that the molecular key to emotional tearing may be a molecule called Neurotrophic Growth Factor, or NGF for short. Originally discovered as a protein that facilitates the development and survival of neurons, NGF has a healing effect in our eyes as well as a role in the regulation of mood.[11] For Provine, although the routes and reactions by which this may happen are still not evident, the presence of NGF in tears and its access to the nervous system make it a good candidate to be what gives the salty fluid of tears its emotional texture.

What fascinates me, but I believe still awaits explanation, is first the threshold of intensity in the causing event that triggers the crying reaction, and second the question of what makes certain people more prone to crying than others. Tears are connected to sadness and desperation, but, of course, we sometimes cry out of joy and happiness, out of emotions that bring gratification and recognition instead of depriving us of something. In both cases, what makes tears overflow is unknown. We are all familiar with the languid moment when tears well up in our eyes. It feels like an evening tide that rises suddenly under the feet. There may be periods when crying is an uncontrolled inundation. Tears push heavily on our doors and flood the chambers of our being, unsolicited. But there are other times when tears refuse to come, even when we really would like them to, and we are left in a dry desert. Even when we can't tell why we are crying, tears are the bringer of some important message that is hidden somewhere in the secrets of our unconscious.

## Is grief similar to physical pain?

Grief over loss and other shadings of emotional aching are often articulated in the language of physical pain. When hit by disappointment, rejection or damage to relational bonds, we say we are hurt, that someone or something has caused suffering by the infliction of wounds, shallow or deep. We feel beaten up and crushed. We are left with scars.

The relationship between physical and emotional pain goes beyond semantics. Physical pain and emotional pain – the pain we suffer when our social and emotional bonds are broken – may share some of their underlying neural mechanisms.[12] From an evolutionary perspective this would make sense. The system mediating the experience of physical pain has older roots upon which the system for emotional pain may have developed. We experience physical pain so that we can avoid hurtful experiences.

Grief is more like interest on an emotional debt. It is the inevitable costly price we pay for attaching to others.

The causes of physical pain and grief are different, yet they carry similar effects, at least at the level of neurons. Emotions are aroused by events, or by thoughts and images that remind us of those events. In all cases, something moves under the skin and our bodies process the change. When I heard of my grandfather's death, I was still in bed in my London flat. The call came unusually early in the morning. I knew that he had been very ill for the past several days. When I heard the telephone ring, I was sure the call came from home and already knew what I would be hearing. Even though I had prepared myself for his departure, it was only after listening to the voice of my sister at the other end of the line announcing Nonno's death that the cascade of grief began. The physical pain sensed when we stub a toe or hit a wall is the effect of a *collision* that damages our tissue. The pain of loss or the breaking of an emotional bond, on the other hand, is the consequence of a physical *separation*. Something departs from our surroundings and from our lives. All the same, its disappearance hurts us, causing pain just as would colliding with the wall. However, unlike a cut or a bruise, it is the absence of a loved one that hurts and leaves a mark on us, and that can even be harder and slower to heal. We must get used to the idea that we'll no longer be able to see or touch them. It is an incredible effort to accustom ourselves to the fact that a person no longer exists. We must *unlearn* their physical presence and their dwelling-place in our emotional universe. All our senses must adjust. We conjure the lost one up by weaving again the neural networks that used to make us perceive them. As expressed in the lines by Borges I have used as an epigraph to this chapter, the absence of the lost one surrounds us and the experience can be as suffocating as having a rope tighten around the neck.

Clues to neural commonalities shared between the effects of physical and emotional pain have come from several sources. Research studies on palliative drugs are one such source. Opiates, such as morphine, work to sedate and reduce excruciating physical pain. They also work to reduce the pain resulting from separation. As I mentioned briefly earlier, although animals don't shed tears, they do protest on being separated from their mothers or caregivers, by emitting shrieking vocalizations. It has been shown that if you give opiates to young animals (of various mammalian species) separated from their mothers, their vocalizations of protest and distress diminish.[13] Another set of data linking physical and social pain comes from neuroanatomical and imaging studies and involves the dorsal anterior cingulate cortex, a large structure in the middle part of the frontal lobe. For a long time, the dACC has been linked to physical pain. For instance, creating lesions of the dACC – a surgical operation called cingulotomy – has been used as an effective treatment of chronic pain disorders. Recently, an involvement of the dACC in the modulation of social and emotional pain has also been tested. The neuroscientist Naomi Eisenberger and her colleagues explored the experience of social pain by measuring the neural activity during an experience of social exclusion simulated in a brain-imaging experiment.[14] The simulation involved a ball-tossing game. The participants lying in the brain scanner were told they were playing with two other people via an internet connection. In truth, they were alone; the other players were computer-generated images of people tossing the ball. In one round of the game, the person in the brain scanner would be included in the game and passed the ball by one of the two players. In another round, they would be excluded. The brain region found to have a stronger oxygen flow during the period of rejection and exclusion from the game than in the period of inclusion was indeed the dACC (and another region called the periaqueductal grey area).

Similar results were obtained in a study that specifically investigated grief. The experiment consisted in showing a group of bereaved women pictures of those they had lost. The pictures were matched by words related to loss or grief that had been taken from the participant's own account of the death.[15] When the women's painful reactions to the pictures of their loved ones were compared with their reactions to pictures of strangers, the brain areas that were involved in the emotionally painful reactions were regions known to be linked to physical pain.

The similarities between the brain regions identifiable across different kinds of pain are definitely interesting. But of course this doesn't mean that grief may be easily reified within precise precincts of the brain. My cautionary remarks in chapter 2 about the limitations of attempts to identify the neural locus of guilt hold just as true for grief, which, as I will explain later, is a variegated concept with a long history.

## Good grief

Though bewildering and, at times, debilitating, grief is not intuitively regarded as an illness. Yet in today's society, bereavement may attract medical attention and be seen as a divergence from normality. This has to do with how, in certain cases, the psychiatric category of depression has turned ordinary sadness into an illness. To understand what I mean by that, we need to briefly unearth the history of depression and go back to the DSM.

As I briefly mentioned in chapter 3, the guidelines for the classification of psychiatric disorders were not derived from knowledge of their aetiology – a term used in medicine to indicate the causes of a disease – but from the commonalities or differences in the symptoms they manifested. In the 1950s, nobody had a definite idea of what caused a depressive mood, but they more or less knew what it looked like when they encountered it in a patient.

When the first edition of the DSM was released in 1952, it contained around one hundred items. The second edition, published in 1968, contained almost twice as many. There were about three hundred mental ailments listed just over a decade later in the third edition (1980). The fourth edition (current to May 2013), first published in 1994 and then revised in 2000, lists in total almost four hundred disorders. Do the maths: the number of recognized psychological ailments increased fourfold in the fifty years following publication of the first edition, with a hundred or so added in each successive edition. This is an impressive escalation and it doesn't seem to be relenting.

Already in use as a term describing low moods in the mid nineteenth century, 'depression' has appeared as a clinical term in all the DSM volumes under different disguises.[16] The 2000 edition of the DSM (DSM-IV TR) makes the main distinction between bipolar disorder, characterized by drastic mood swings, and the category of major depressive disorder (or MDD), which typically refers to an enduring low mood and is what we commonly refer to as depression today. The current list of clinical criteria for a diagnosis of MDD includes as symptoms intense sadness or feelings of emptiness, insomnia, decreased appetite and loss of weight, fatigue and loss of energy, diminished interest or pleasure in usual activities, difficulty concentrating on regular tasks, as well as feelings of worthlessness or inappropriate guilt, and recurrent thoughts of death, suicidal ideation or attempts. Importantly, in order for a diagnosis to be made, such symptoms – at least five of them, of which two are required to be sadness and the loss of interest in pleasure – must occur 'most of the day, nearly every day for at least two weeks'.

If you have experienced grief yourself or witnessed it in others, you will have observed that most bereaved human beings suffer from most, if not all, of the above symptoms, in more or less intense shadings. Anyone who has recently lost a partner, a friend or a relative will experience an overwhelming period of adaptation to that loss. Actually, it would be rather surprising if they didn't.

In his influential essay 'Mourning and Melancholia', Freud explains the commonalities between what we would nowadays call grief and depression. What the two have in common is an enforced separation from someone or something we grant our attention and love. We could say that the parting is a theft of an emotional investment. In the case of grief, the separation is caused by an actual death. In the case of depression, the separation is unconscious and cannot be physically perceived. It may involve the loss of something, a reaction to being 'slighted', 'neglected', an ambivalent emotion that is starved of its fulfilment. In other words, grief comes from without, depression from within. But, in both cases, such separation procures pain. In both cases, the individual retreats from reality, turns inward, loses interest in the outside world. Those who eventually recover from grief then slowly adapt to reality and accept the loss. Depressed people continue to isolate themselves, they are prone to self-criticism and self-reproach, and lose self-esteem. So, grief is justified and liberating, whereas depression can get out of control. Freud clearly states that 'although mourning involves grave departures from the normal attitude to life, it never occurs to us to regard it as a pathological condition and refer it to medical treatment. We rely on it being overcome after a certain lapse of time, and we look upon any interference with it as useless or even harmful.'[17]

Indeed, the 2000 DSM edition did not list grief as a clinical disorder. Bereavement is excluded as a disorder because the authors recognize that depressive symptoms are to be expected in recently bereaved individuals. Already in the introductory pages of the manual, where the authors provide a general definition of mental disorder, they say that for a condition to be granted clinical status it 'must not be merely an expectable and culturally sanctioned response to a particular event, for example, the death of a loved one'.[18]

A fifth edition of the DSM, updated and restructured, has recently been prepared (published in May 2013). A particularly worrying change has been introduced: the task-force that worked on the new version of the manual scrapped the exclusion of bereavement.[19] In simple terms, this means that a grieving person whose symptoms of depression persist for a period longer than two weeks is in principle entitled to earn a mental illness diagnosis. One of the arguments put forward by the proponents of this change is precisely the fact that, at the level of symptoms, there is little if no difference between those who grieve and those who develop depression for reasons other than someone's death.[20]

The question then arises: is the biology behind the symptoms different in the two circumstances? A few researchers are trying to identify symptomatic and biological factors that could justify the creation of a new dedicated category named *prolonged grief disorder* (PGD) or *complicated grief* (CG), thus differentiating normal grief from a form of unresolved grief which deteriorates into an incapacitating condition that parallels severe cases of depression.[21]

All in all, the proposal comes with the best of intentions. Doctors have no specific interest or desire to increase the toll of psychiatric disorders in the world by over-diagnosing them. The world prevalence of major depression currently stands at approximately 10 per cent of the population.[22] That means that one in ten individuals that you see walking on the street may be depressed. In the United States, approximately 2.5 million deaths occur every year.[23] If, on average, each death leaves four or five bereaved survivors, then about ten million people each year could potentially be diagnosed with PGD. The main argument in favour of the introduction of prolonged grief disorder is that it would become legitimate for doctors to detect it and treat it swiftly to avoid the onset of a much more complicated illness (and as a practical consequence, especially in the US, insurance companies would be more likely to reimburse its care).

The change made in the new DSM edition inevitably bears unwanted consequences. Leading psychiatrist Allen Frances, who was also the chair of the task-force behind DSM-IV, has several times warned against the creation of a new category for grief.[24] Establishing boundaries of duration to distinguish normal grieving from a form of grieving that demands special attention and dedicated treatment is bound to generate a large number of false positives. Nobody can really tell what a normal duration of grief ought to be. Two weeks is definitely too short a time to conclude a season of sorrow for the death of a loved one. Most of the people whom I have seen cope with grief take much longer than that. And there doesn't seem to be empirical evidence proving that all those who take longer than two weeks to recover from the gripping symptoms of grief will end up being incapacitated by the loss. Depending on the life circumstances of the bereaved – health, work and financial conditions, past experiences of grief and other difficult life experiences – the individual variation in the duration of grief is enormous, just as is the variation of symptoms in depressions not caused by the actual loss of a loved one.[25]

Grief is also articulated by factors such as culture. Different mourning rituals and traditions encourage different lengths of seasons of grief which, in response to the disorienting experience of loss, help the bereaved by providing guidance and structure on how to cope with it. If I told Nonna that her grieving season might be mistaken for something abnormal, she would probably take offence. Speaking of somebody grieving more or less intensely creates a hierarchy of emotions that undermines their value. The categorization of grief would turn it into a commodity.

The writer Julian Barnes once said that mourning 'hurts as much as it is worth'.[26] Mourning is painful, but it is necessary in order to deal with the loss. Indeed, the most dangerous, perhaps unintended, consequence of this move is that normal grief, an entirely expected reaction to loss, may be wrongly branded an unwanted problem. A new category for grief is just a label. But with its introduction would come millions of patients who, before the label existed, would not have been considered candidates for medical attention.

## By any other name . . .

In 1953, only a year after the publication of the first edition of the DSM, another very important volume made its debut in the world. It was the posthumous publication of a charming, mysterious and out of the ordinary Austrian philosopher who taught at the University of Cambridge.[27] The book in question is *Philosophical Investigations* and the philosopher Ludwig Wittgenstein (1889–1951). The Austrian thinker was obsessed with language.

For Wittgenstein, language is the staple of our social lives, but also underlies most misunderstandings and disagreements.

Wittgenstein was firmly convinced that the meaning of words is not an inflexible correspondence between an arbitrary string of letters and an object or entity in the world. On the contrary, words assume meaning according to the use which we make of them in the outer world. Wittgenstein called the use we make of words their public aspect and believed it had more influence than the private one. For him, the grammar of a language was not about how to put together a sentence correctly, minding rules of syntax and orthography, but about the set of rules or customs attached to the use and meaning of a word. He adopted the term language games to describe everyday social contexts in which words were employed for particular purposes and according to particular rules.

His most famous example is indeed the word *game*. We have board-games, card-games, ball-games, Olympic games, war-games, etc. All such words have 'game' in common, but they all actually mean something different.

The practice of psychiatric diagnosis consists exactly in associating a name with a list of symptoms, a set of behavioural patterns that are supposed to give meaning to a disease. In turn, each diagnostic term implies the existence of some disease entity, which we know conceals a complex biological scaffolding, the structure of which, however, we are only starting to comprehend. This is true of depression – in all the various designations in which it has appeared over time. The proposed new category of prolonged grief disorder therefore purports to correspond to something specific, distinguishable both from major depression and from what would be 'normal' grief.

Wittgenstein was neither a medical doctor nor a scientist, but had an interest in psychiatry. The language problems Wittgenstein formulated for everyday words hold true for the categories of the DSM, which, in one way or another, enter the everyday language of doctors, researchers and patients alike, and even the language of the media and public discourse. Major depression, bipolar disorder and all the other categories pervade everyday talk and work as terms by which individuals define themselves and their condition. Knowledge of the existence of a diagnosis, a name for a mental illness, as well as a biological description, is often a comforting discovery for patients, one that erases a sense of self-blame for being ill.

Although the impact of Wittgenstein's work and legacy has-been greatest on logic and the philosophy of language in general, the contribution of his thoughts to the field of emotions was far from marginal. We need to take a step back and examine it.

As I have emphasized several times in this book, one of the current prevailing concepts in research on emotion is the distinction between emotions and feelings. This distinction is often used to clarify that emotions are spontaneous bodily reactions to events and circumstances, and feelings are internal, subjective and private states, fruit of introspection and awareness of those emotional states, and hence not accessible to others. As a consequence, those around us can only deduce what we feel, and achieve an approximate interpretation of our internal states. So far, so good.

Wittgenstein recognized and endorsed the idea that emotions are immediate visible manifestations.[28] 'Don't think, but look!' he urged, implying that the bodily expression of an emotion communicates much more than can a description of it and needs little learning or interpretation.[29] Behaviour and what was observable to the naked eye mattered to him a great deal and he acknowledged the power of our bodies to effectively communicate emotions to each other: 'Grief, one would like to say, is personified in the face. This is essential to what we call "emotion".'[30] Facial, vocal and other bodily expressions were all valid manifestations of emotion, while language was a secondary, yet determining attribute. If you leaf through Wittgenstein's works, you will find some pages adorned with drawings he made to allude to emotional expressions and to aid his arguments – among them the following passage:

> If I were a good draughtsman, I could convey an innumerable number of expressions by four strokes –
>
>   
>
> [ . . . ] Doing this, our descriptions would be much more flexible and various than they are as expressed by adjectives.[31]

In fact, they are not dissimilar to today's emoticons.[32]

What Wittgenstein didn't believe was that introspection could reliably extrapolate the essence of mental states. In Wittgenstein's philosophical grammar, the terms we use to denote emotions such as grief are not direct correspondences to our inner states. He didn't mean that we can't build inner feelings or that introspection doesn't work. We certainly do have subjective experiences of our emotions. However, Wittgenstein thought that we do not learn how to identify our emotions solely through the inner experience, but through the language we use to describe them – in addition to our expressions. Just as in the case of the word 'game', without a public set of criteria to describe emotions, there would be no way to understand what we mean by them, let alone judge what others are feeling. The way we describe emotions depends on the available public language of emotions and also on the situation as well as the historical context in which they arise.

## My grief is not yours

It is ironic that the two publications were issued at pretty much the same time: the American Psychiatric Association's manual that prescribed the language and categories by which mental pathologies and, by extension, emotions such as grief were to be labelled, and Wittgenstein's reflections on the impact that language and words have on the way we understand our lives and interact with each other. Wittgenstein did not live long enough to witness the publication of the DSM, or to marvel at the scientific advances of the second half of the twentieth century. He had no knowledge of the structure of DNA – the publication reporting its discovery was issued two years after Wittgenstein's death – or of the roles we attribute today to the amygdala, the prefrontal cortex and neurotransmitters such as norepinephrine and serotonin. Serotonin had been isolated in 1933, but it was only after the philosopher's death that it was associated with emotional states.

Wittgenstein, however, must have had his own notion and experience of grief. It doesn't matter too much whether Wittgenstein believed in the existence of indefinable inner feelings and in the explicative power of introspection. After all, as yet nobody knows the exact composition of these inner feelings and consciousness, or for that matter how we can measure them. And not everyone agrees that such feelings ever will be measured with exactitude. If Wittgenstein were still alive today, sixty years after the publication of the Investigations, it would be fascinating to ask his opinion on the current state of psychiatry and on the most recent neuroscience developments. Would he show curiosity at the whole neuroscience enterprise? He would probably be puzzled by and perhaps cringe at the very idea of defining a single diagnostic category – prolonged grief disorder – to encompass the wide and complex spectrum of emotional state that we label grief. He would also probably have doubts on what really lies behind that name.

Research in psychiatric neuroscience is heading towards the identification of biomarkers. These are measurable biological values that work as proof of some distinct change in the body. For instance, high levels of gonadotropin in a woman's urine are the biomarker of her pregnancy. Insulin level is a good indicator of whether someone has diabetes. In the case of mental illness, biomarkers would indicate dysfunction in the neurochemistry of mental states, thereby facilitating the diagnosis and the treatment choice for depression or complicated grief. Over the decades of neurological and molecular research on depression biomarkers have varied widely. To take some examples: levels of cortisol – the hormone involved in an organism's stress response – appear to be higher in depressed individuals, especially during the early hours of the day; certain alterations in brain morphology or changes in brain activity, detectable through brain-imaging techniques, have been shown to indicate depression;[33] depressed patients have also been found in general to have decreased blood flow in the frontal part of the brain. Researchers and the authors of DSM-V are eager to pinpoint as many reliable and precise biomarkers as possible and include them as diagnostic criteria.[34] This search is an exciting undertaking, because it leans towards a refinement in diagnosis. However, it is also an extremely challenging one because of the vast diversity within any one psychiatric disorder, both at the level of symptoms and at the level of the biology underlying them. There are so many variables involved and it is unlikely that one biological measure could suffice for a diagnosis.

Even if, as human beings, we may universally experience grief and share a few of its common biological components, in its detail it varies greatly from individual to individual. Again, some of the variation stems from cultural rituals. I see my grandmother's bereavement in the flowers she buys and the candles she lights, in the modest dark clothing she wears, in her dusting of the telescope, in the pauses she makes when she speaks about Granddad, and in the fish soup she cooks in his honour for family gatherings, because that is what grief looks like in her: none of which can be conveyed by the mere word 'grief' – or prolonged grief disorder, for that matter.

The psychiatrist Ronald Pies has used Wittgenstein's concept of 'family resemblances' to express how difficult it is to describe mental states, especially in the context of psychiatry.[35] Wittgenstein suggested that when we look at a family portrait, most probably there will be no one feature shared by all members of the family. However, if we inspect the picture closely, a few resemblances will become apparent. Five members might all have freckles, and three of those five might also have blue eyes, as do several non-freckled people in the portrait; three other members of the family may be the same height. Taken in combination, such features are indications that the people pictured are all related, even though there is no single feature present in all. The same applies to psychiatric disorders. No two people sharing a diagnosis of depression are exactly the same. And, for sure, where grief is concerned one size does not fit all either. There will always be huge variation between individuals in their experience of grief. Equally, the course of each individual's path to recovery will be strictly personal. Categories in psychiatry are black or white diagnoses: you either have the disorder or you don't. But when one is making a detailed study of symptoms, or looking for the neurological or genetic factors underlying a disorder, methods of assessment and measuring systems that account for diversity will be more helpful.

## The molecule of sadness

One of the most pervading popular narratives about depression is that it is the outcome of a chemical imbalance – more specifically, a decline in the level of neurotransmitters in the brain.

Neurotransmitters, the molecules that relay messages between neurons in the brain, have entered everyday vocabulary. Here are just a few examples: we find ourselves associating the pleasure gained from sport with the release of endorphins. We speak of an 'adrenalin high' when it keeps us alert and insomniac after a test, a performance or an important meeting. Occasionally we mention the hormone cortisol to describe or justify our levels of stress. But if there is one molecule that has truly become a household term, a topic for tube or dinner conversations, a recurrent word in science magazine titles, it must be the neurotransmitter serotonin. I have often heard and cringed at statements like: 'My serotonin levels must be low today' or 'That man needs to boost his serotonin'.

Serotonin has a simple molecular structure (Fig. 11): twenty-five neatly arranged atoms. It is heralded too easily as *the* molecule of happiness, used as a sloppy shorthand term for the status of our brain and for our well-being. Serotonin has become so

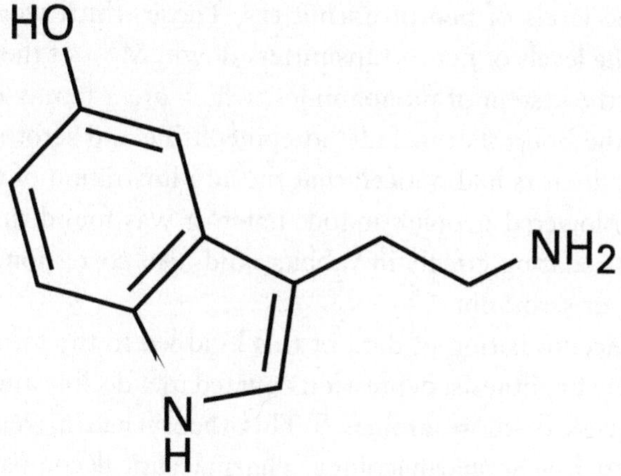

Fig. 11 Molecular structure of serotonin

popular that it is possible to spot its molecular structure printed on mugs, T-shirts, postcards, moulded as jewellery and even tattooed in praise of its properties as a mood-lifter.

Serotonin is not exclusively present in the brain. Approximately 90 per cent of the entire amount of serotonin in the body is in fact stored in the intestines. There, it facilitates gut movements through the regulation of the expansion and contraction of blood vessels, and is also involved in the function of platelets, the blood cells that promote the coagulation of blood and the closure of a wound. Only the remaining 10 per cent of serotonin accomplishes its other duty, that of neurotransmitter in the brain, where it is produced by dedicated serotonergic neurons, mostly in a structure called the raphe nuclei. These are located in a central part of the brain, along the midline above the brainstem, and have neuronal connections extending to almost every part of the central nervous system.

The discovery that mood may correspond to a neurochemical imbalance in the brain is to be traced back to the 1950s. It was based on a series of unexpected observations, some of which were made in animals, that a few drugs interfered with mood. Some drugs improved it, some worsened it. Those which improved it raised the levels of neurotransmitters. Those which worsened it pushed the levels of neurotransmitters down. Most of these drugs targeted the system of monoamines, which are a family of molecules in the brain that include norepinephrine and serotonin. For instance, doctors had noticed that the administration of the drug reserpine lowered people's mood. Later, it was found that reserpine had sedating effects in rabbits, and also corresponded to a lowering of serotonin.[36]

The accumulation of data of this kind led to the formulation of a simple hypothesis: depression equated to a decline and elation to an excess of those amines.[37] This theory had a tremendous impact on psychopharmacology. Pharmaceutical companies began to synthesize drugs that worked to increase the presence of neurotransmitters.

In order to understand how drugs really affect serotonin, let's brush up on some of the basics of neurochemistry.

The hundred billion neuronal cells that make up your brain do nothing but talk to each other. Remarkably, the communication goes on without the neurons having to touch. The 'language' in which messages are conveyed consists of sequences of neurotransmitter molecules and the dialogue between cells takes place across a tiny empty space called the synapse, the point of encounter between neurons. Picture this space as a channel separating two shores, and the neurochemical shuttling of information between one neuron and the other as an old-fashioned exchange of letters, with neurotransmitters like serotonin being reliable postmen on boats – my granddad could have been one. Whenever a neuron needs to communicate a message, it launches the relevant neurotransmitter into the channel. Waiting on the opposite shore are receptors, the recipients of the letter. There are at least fifteen different types of receptor that can receive the message from serotonin, each with a different role in coordinating various aspects of mood (in chapter 3, for instance, I mentioned the serotonin receptor 1A that contributes to keeping anxiety at bay via its inhibitory function). The delivery system is extremely accurate and, as it were, confidential: the message can only be read by its intended recipient – that is to say, serotonin binds only to serotonin receptors. The recipients do not keep the message. After having been opened and read, the letters are sent back into the channel, the synaptic cleft. In the meantime, the sending neuron has dispatched more letters across the channel, so at some point there may be too many boats floating around – too much serotonin. When that happens, those extra boats must be cleared because the whole system strives for an equilibrium.

There are two main strategies by which the serotonin is cleared from the channel to keep the right balance. The first is through the action of enzymes that degrade it. Maintaining the marine metaphor, think of such enzymes as if they were sharks that chew up the floating serotonin. One such shark is the infamous MAOA, a major serotonin degrader. When drugs were first developed specifically to maintain high levels of serotonin there was a class that were, in fact, inhibitors of MAOA.

The second strategy is to clear the synaptic cleft of serotonin by sending it back to where it came from, a sort of conscientious paper recycling. This is done through the action of dams or embankments on the neuron of origin that suck up whatever neurotransmitters are present in excess.[38] Serotonin has one such embankment dedicated to its 're-uptake': it is a large protein on the outer walls of the neuron, known as the serotonin transporter. Too soon this became a target for drug treatment in the quest for enhanced levels of serotonin. A new class of medications exploded on to the scene, the selective serotonin re-uptake inhibitors (or SSRIs). Prozac was born, and following its incredible commercial success a host of similar drugs were introduced. Medications like Prozac, Zoloft, Sertraline or Paxil all work by inhibiting the serotonin transporter and trying to increase the amount of serotonin available for its receptors – the recipients on the other shore.

Ever since their introduction into the pharmaceutical market, SSRIs have enjoyed an impressive career, at least from an economic perspective. Over thirty antidepressants have been released.

In the United States alone, one of the countries with the highest consumption, in 2011 the number of antidepressant prescriptions went beyond 250 million, over a hundred million more than in 2001.[39] These remarkably high figures correspond to sales equivalent to $25 billion.

Yet this economic success is not matched by improvement in the population's overall mental health, if we consider the high incidence of depression in the world. In Europe, the largest toll out of the total burden of illness is attributable to psychiatric disorders.[40]

That a deficiency in serotonin is the cause for a low mood is not a definitively ascertained hypothesis and results from ongoing work aimed at resolving this question remain contradictory. Apart from some of the general and opening stages in the chain of reactions that I have just described, the exact molecular mechanism by which common antidepressants work is not fully understood. We have a fairly neat picture of how serotonin accomplishes its role in the synapse, but our knowledge of exactly how the mechanism then translates the message into cell events and mood changes and of what makes the drugs effective is far from complete. Regardless of that, for a couple of decades pharmaceutical companies have used a simple and easy-to-remember slogan telling us that when it comes to serotonin, the more of it you have, the better you feel. Direct-to-consumer advertisements continue to use this simplistic equation to 'explain' to a lay, non-expert audience what for the neuroscientist is still an unresolved, complex scientific question.[41]

In 2012, the pharmaceutical company GlaxoSmithKline was heavily fined for having bribed doctors to continue to endorse and prescribe to children and teenagers the antidepressant Paxil (paroxetine), even though trials had shown it was only effective in adults, and its use in groups of younger individuals had been linked to risk of suicide.[42]

In February 2008, producers and consumers of antidepressants were taken aback when a scientific report called into question the efficacy of antidepressants. The report surveyed a large set of data from clinical trials – including unpublished data – submitted to the US Food and Drug Administration to obtain approval for the most common SSRIs. The data were comparisons between the effects of pills and placebo on depression patients. In a nutshell, the report concluded that the prescription drugs were no better than placebo in treating those patients who manifested only mild-to-moderate depression.[43] The results were greeted with dismay especially among those for whom the drugs in question were essential bearers of comfort and the only support for a functional existence. It sounded as if they had been taking medications which in fact were no better than a sugar pill. Indeed, in the past five years or so, some of the large pharmaceutical companies have reduced their investment in mental health pharmacology and are looking for new prospects.[44]

Exactly where on the scale of depressive symptoms it becomes appropriate to prescribe medication for a patient remains a controversial question just as is the choice between diagnosing or not. Not for one minute do I want to say that antidepressants never work or that they should not be prescribed. It's obvious that some people greatly benefit from them. However, consumption figures indicate that they are prescribed way too easily – and a specific diagnosis for grief is unlikely to counteract this trend. What we should bear in mind about depression is that there is more to it than the metabolism of serotonin. It is worth pursuing the search for new medications involving different molecules and other neurochemical pathways.[45]

## Treatments, old and new

When I walk on the beach along the Sicilian coast where I grew up, especially when no one is around, I often think of who might have paced the same shore millennia ago. This corner of the world has been a crossroads of many and great civilizations, the stage for many wars, but also the cradle of great ideas and magnificent art. Not far from where I walk must have strode Archimedes, the mathematician and original thinker most famous for his exclamation 'Eureka!', or 'I have found it!' – which is now inevitably brought out again each time someone has a great idea.

In the fifth century BC an illustrious visitor reached these shores on a trip from Athens. This was Hippocrates, the renowned physician who is considered to be the founding father of medicine and who surely knew how to cure a bout of sadness.

If today grief, sadness and depression are articulated in terms of neurotransmitters and their imbalance in the brain, back then, they were the outcome of a different kind of imbalance. Hippocrates understood moods and behaviour in terms of *humours*. Humour is a word of Greek origin literally meaning fluid. The general idea was that inside our bodies streamed a combination of four fluids, each with different properties, that worked to make up our health, both physical and mental.[46] These four humours were phlegm, blood, yellow bile (or choler) and black bile (or melancholy). Where did they originate? Descendants of the universal cosmic elements – water, air, fire and earth, respectively – the humours were believed to be some side-product of the digestive operations in the stomach, processed in the liver and further refined in the bloodstream, and bathing all parts of the body, including the brain. Hippocrates granted the brain a primary role in determining health, modulating sensations, thought and emotion:

> . . . the source of our pleasure, merriment, laughter and amusement, as of our grief, pain, anxiety and tears, is none other than the brain. It is specially the organ which enables us to think, see and hear, and to distinguish the ugly and the beautiful, the bad and the good, pleasant and unpleasant . . . it is the brain too which is the seat of madness and delirium, of the fears and frights which assail us, often by night, but sometimes even by day.[47]

The exact appearance of the humours was not discernible, but they were to be found within visible fluids and discharges of the body. The humour blood was indeed part of the blood circulating in arteries and veins. Phlegm was present in the mucus of a runny nose and in tears. Choler hid in pus and vomit. Black bile was posited to be part of clotted blood or dark vomit. For Hippocrates, each person had their own composition of humours and the occurrence of illness was a disruption, an alteration of his or her humoral balance. Hence, treatment consisted in remedies that attempted to restore such balance and permit a return to the original equilibrium, for wherever there was balance, there was health. The degree of concentration of the respective humours and their proportions in a person's internal blend were held responsible for the behaviour, temperament and mood that person manifested. Roughly speaking, an excess of phlegm made a person phlegmatic and peaceful. Too much choler caused irascibility. A glut of blood made people sanguine, that is upbeat and positive. An excess of black bile guaranteed the onset of melancholy.

One of the most fascinating aspects of the humours is that they were purported to be in constant dialogue with and a reflection of the external world. The internal microcosm of the body mirrored the external macrocosm and order of the universe. Hippocrates made the humours match the course of seasons and stages of life. Thus blood corresponded to spring and childhood, choler to summer and youth, black bile to a melancholic autumn and to maturity and phlegm to winter and old age. An individual's humours were sensitive to the environment. External temperature and the seasons influenced the humoral composition. Heat and cold and the consequent conditions of dryness and moistness affected the overall balance of the humours and therefore the resulting mood. So, for instance, it was normal to feel hot and dry and be full of choler in the summer or have an excess of phlegm in the winter, which is cold and moist (Fig. 12). Hippocrates specifies how these imbalances affect the brain:

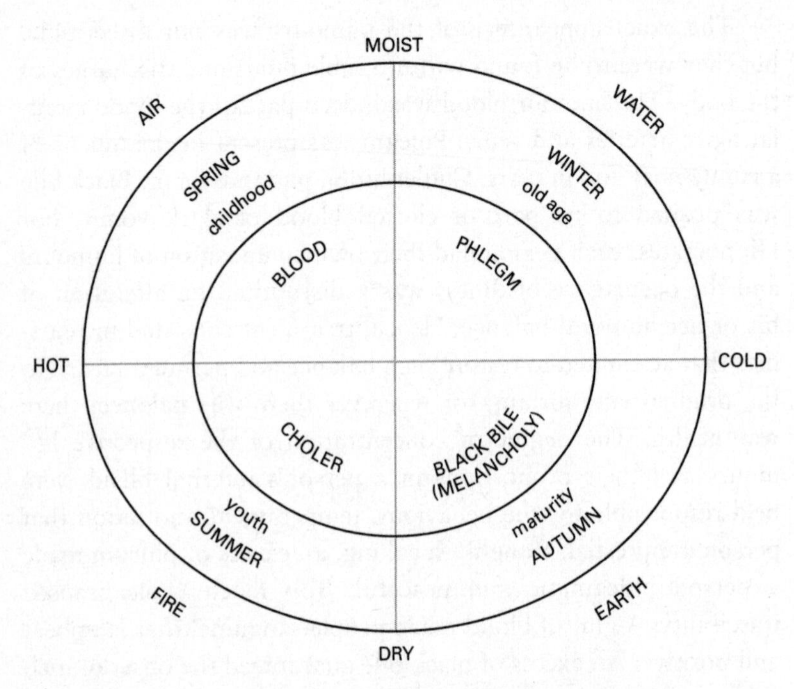


Fig. 12 Schematic view of the four humours and their correspondence to the four elements, seasons and phases of life (diagram adapted from Arikha, 2007)

> The brain may be attacked both by phlegm and by bile and the two types of disorder which result may be distinguished thus: those whose madness results from phlegm are quiet and neither shout nor make a disturbance; those whose madness results from bile shout, play tricks and will not keep still but are always up to some mischief.[48]

Since it was the heat of digestive processes in the stomach that was responsible for producing the humours they were also sensitive and responsive to a person's diet.

Leaning on the authority of Hippocrates, the humours survived as a valid theory for over a thousand years and were relayed from healers to philosophers to doctors at least until the Enlightenment, flourishing among Roman physicians, in Arabic medicine, and in European medicine during the Middle Ages and Renaissance.

Within the humoral framework, grief and sadness belonged to the condition of melancholy, and were, therefore, caused by a glut of black bile. Such excess produced symptoms such as despondency, dejection, tendency to suicide, aversion to food and sleeplessness – a list very similar to the criteria for a contemporary diagnosis of depression and to the symptoms of former versions of depressive illness, such as Freud's melancholia. Some of the ancient medical treatises specifically talk about grief as an emotional reaction provoked by external events such as the separation from a loved one.[49]

In general, such events caused an individual's internal vital heat to subside. The treatises offered specific prescriptions and therapeutic recommendations including bodily regimens that ranged from physical exercise to particular food ingredients. The most important recommendation was to keep the body warm, to restore the heat and fight the cold dryness of the black bile, for instance by taking regular lukewarm baths. But there was also specific food advice. A melancholic was better off with a diet that included lettuce, eggs, fish and ripe fruit. He or she should avoid acidic foods, such as vinegar. The ideal day of a melancholic person would include a routine of walks, exercise, massages with violet oil, as well as sessions of music, poetry and recitals of stories or tales from the lives of sages.[50]

Nowadays the ancient theory of the humours is considered inadequate to address the variations of mood and behaviour we ride in our lives. Nevertheless, today's neurotransmitters and electric impulses are simply what humours used to be thousands of years ago. What the legacy of the humoral theory and the historical recurrence of the melancholic type do is remind us of the fact that the emotions of sadness, grief and melancholy have always existed. Depression, prolonged grief disorder and melancholy are permutations of the same emotion that have just been understood in different terms. I am not saying that we should embrace the humoral theory, nor that we should abandon neuroscientific research into the molecular basis of sadness. However, given some of the problems in the current system of diagnosis, the diversity of symptoms and underlying biological factors within a psychiatric condition, the multiplicity of its possible causes, not to mention the uncertainty about how effective contemporary treatments may be, there is room for a broader approach in treating patients. Especially those who are suffering from grief. Even though Hippocrates considered the brain to be an important centre for an individual's emotions and tempers, his treatments were intended for the whole body and valued the uniqueness of each ailment in every patient.

An editorial in a recent edition of the *Lancet*, in an impassioned plea against the category of PGD and the risk of over-diagnosis and over-medication, stated that doctors facing the treatment of bereaved people 'would do better to offer time, compassion, remembrance and empathy' rather than the more advanced and synthetic therapeutic options that have come to the fore through the swift development of psychopharmacology.[51] This is not far removed from more ancient medical remedies and would conform to one of Hippocrates's tenets in the practice of medicine, which was to 'do no harm' to patients.

## Coda

'The cure for anything is in salt water: sweat, tears or the sea,' wrote Karen Blixen, in *The Deluge at Norderney* (under the pseudonym of Isak Dinesen). There is reassurance to be gained from this statement. We earn reward for every effort exerted. We feel better after the liberating action of a good cry. We can draw strength from the capacious calmness of the sea.

Gazing at the sea is a nurturing activity. Whenever I come to Sicily to visit my grandmother and return to the places where I spent all my summers as a child, I regain comfort and energy. I travel south to the very tip of the island to marvel at the horizon, planning my journey so that I arrive at sunset. When as a child I learnt the basics of geography, and about the movements of the earth, the moon and the universe, I found it simply magic that the sun, which on the east coast always rose from the sea and disappeared down behind the hills, could set over the sea if I just walked around the tip of the island towards the west – one of the advantages of living on an island. I wanted to come here every day because it felt as if I was turning the world upside down, and I rejoiced in the ritual and in the change of perspective.

Sunsets are hypnotic and have always been especially conducive to the melancholic mood. Melancholy assumes its most agreeable form at dusk. It belongs to the evening. Light is a brush, gently painting everything with a crepuscular tint. Somebody said looking west is like searching for immortality. When I stare into the depth of the horizon, I am reminded of my granddad and I look for the wake of his boat. The intrinsic and most vigorous, wicked quality of death is its irreversibility. Like candles, life burns in one direction only, until there is nothing left.

There is a poem by Robert Pinsky I found at a friend's house.[52] It goes like this:

> You can't say nobody ever really dies: of course they do . . .
> But the odd thing is, the person still makes a shape distinct and present in mind
> As an object in the hand. The presence in the absence: it isn't comfort, it's grief

Sadly, when my grandfather passed away, my grandmother lost the man who, when alive, was without any doubt the person best at giving her comfort whenever she was sad. Now she needs to make Granddad live again in her memory and, through those images, occupy empty spaces that are just too wide to be filled. And that is what I do too, with Granddad and with the other people I have lost.

# 5

# Empathy: The Truth Behind the Curtains

Those who see any difference between soul and body have neither

Oscar Wilde

So I wish you first a sense of theatre;
only those who love illusion
and know it will go far

W. H. Auden

The lights have slowly come down, as the bell rings for the third time.

'Please take your seats, and remember to switch off your mobile phones,' we hear from a kind recorded voice. 'The show is about to start.'

Next you can only hear the noise of people moving in their seats to reach the most comfortable position and be prepared. A few whispers, the last hissing sounds before the show begins.

Everyone is holding their breath. Theatre is a ritual, one of birth and change. Each performance buds and arranges itself into something new every night, even though it is the same play.

I am here to see my friend Ben Crystal on stage. He is going to be Hamlet. Right now, he is probably waiting in the shadow of the wings. When I go to see Ben act, I always wonder what he is up to during the moments immediately before entering the first scene.

Is he pacing up and down impatiently? Is he wrestling with his memory or murmuring to himself the unruly song of a few knotty lines? Will he see me sitting in the second row?

If, for the audience, the start of a show marks the entry into a new dimension, for an actor stepping into the light must be a rite of passage, a crossing between entire worlds. Depending on Ben's state of mind, that first step on the boards must feel one day like a feather, on another like a stone – and I wonder if the latter is more congenial for playing Hamlet. Melancholy is Hamlet's quintessence, the source of both his craftiness and his misery. But in either case, entering the stage for Ben must be akin to throwing an anchor that will moor him to his element. Acting is his second nature.

When he emerges, everyone's attention is directed at him. 'A little more than kin, and less than kind'. The first line is crisp and reverberates far.

A passage in the second act always grips me for its intensity, and boldly uncovers the true core of acting and theatre. Hamlet has learnt from the ghost of his dead father that he was killed by his brother Claudius, Hamlet's uncle. Hamlet is bewildered. His grief is shot through with outrage and indignation. Hamlet aches. However, he is crushed by his own incapacity to exact revenge. Hamlet is arranging with a cast of players a performance of *The Murder of Gonzago* – with the addition of a few lines written by himself – to mirror the death of his father and test Claudius's reaction to the play as proof of his guilt. He prompts one of the players to recite the speech of Hecuba mourning the death of her husband Priam, the King of Troy. The player's impassioned delivery leaves Hamlet awestruck. How can an actor's fictional emotions be so powerful and, by comparison, Hamlet's real sorrow so vulnerable and defenceless?

*What's Hecuba to him, or he to Hecuba?* asks Hamlet. How is it possible that an actor only needs to imagine grief for his face to pale, his whole appearance to turn sombre, his eyes to shed tears and his voice to break? And all for Hecuba, a woman so distant in time and space? What would an actor do if he happened to have Hamlet's reasons for grief? His feeling would be amplified, Hamlet suggests.

Yet Hamlet can't seem to master his own emotions sufficiently to avenge the death of his father.

As I listen to the man in front of me lament his solitary feebleness, a singular, reliable and generous relay of sentiments takes place. Embodied in Ben's minutiae of enactment, carried and delivered word for word through the acrobatics of Ben's voice, that song of desperation travels across the footlights and invades me. Even if I am sitting motionless, something stirs inside me. I feel the blow. By one degree of separation, I at once participate in Hecuba, the player and Hamlet's grief.

Imperceptibly, I stop seeing Ben and see only the Danish Prince.

In those hypnotic instants, I forget where I am. It is in that state of reverie that I wish those moments might last for ever, that a performance might never end.

## A kind of magic

Anyone still maintaining dualist notions of the separation of mind and body is bound to forsake them in front of a stage, when the velvet curtains are drawn back. Watching a theatrical performance one recognizes the harmonious integration of body, intellect and whatever it is that we call consciousness and feelings.

In the concluding pages of his treatise *The Expression of Emotions in Man and Animals*, Darwin acknowledged the power of theatre to evoke emotions: 'even the simulation of an emotion tends to arouse it in our minds'.[1] To back up this assertion, he recalls Hamlet's awe at the player's ability to manufacture emotions.

Darwin's detailed and vivid descriptions of facial expressions and their corresponding emotions could well constitute a rich resource for actors. Quotes from Shakespeare's plays are used by Darwin as supporting proofs of his own observations. He praises the Bard as being an 'excellent judge' of emotions, and as a man with 'wonderful knowledge of the human mind'. When Darwin describes the emotion of fear, he explicitly cites Brutus's reaction to seeing the ghost of Caesar: 'Art thou some god, some angel, or some devil, that mak'st my blood cold and my hair to stare?'[2] In search of support for his observations on rage, Darwin quotes Henry V's battle speech to his soldiers, when he urges them to 'stiffen the sinews and summon up the blood . . . set the teeth and stretch the nostril wide'.[3] When writing about shrugging, he mentions Shylock from *The Merchant of Venice*.[4] Theatre is definitely a prism through which light is scattered in a whole rainbow of emotions.

But how does theatre cast its magic spell? How can a story embodied into stage action have the power to deeply move an audience and stir emotions?

In the darkness of a performance, we participate in an active emotional exchange. We are launched into a story and the plight of its protagonists. We experience the vicissitudes of fictional characters with unique desires and intentions, the realization of which is often conflicted. By doing that, we shed light on our own. By watching on stage a snapshot of the lives of others, we are watching what could happen to us and we are learning about our own world.[5] This gives us the chance to *empathize* with the characters and grasp what they are going through.

The word 'empathy' made its first appearance in the English language in 1909, as a translation of the German '*Einfühlung*', in turn introduced by the German philosopher Robert Vischer, which means 'feeling into'.[6] Vischer first talked about *Einfühlung* referring to the field of psychology of aesthetic experience to remark how an observer perceives a work of art he or she contemplates. In front of a painting, a sculpture or another type of artwork, a viewer empathizes, or fuses, with it – just as I was absorbed into Caravaggio's painting at the gallery in Rome.[7]

Over time, the term empathy was used not only to explain our relationship to inanimate objects, but also to describe how we can instinctively understand other people's mental states.

Empathy lets all kinds of emotions reverberate amongst us. It is the capacity to recognize and identify with what another person is thinking or feeling, and to *react* with a comparable emotional state.[8]

Empathy is the backbone of our social life. Whether in thoughts or acts, it intrinsically demands an interaction with others. It has the power to spread joy, euphoria or laughter, but it also helps mitigate difficult circumstances – for instance, alleviating negative emotions. Anxiety, guilt, sadness, despair are somewhat eased if shared with others. Empathy is like an invisible bond with the power to unite us to other human beings and blur the dividing line between ourselves and them – as in the case of me and Hamlet during Ben's performance.

In this chapter I am going to use theatre as a vehicle to understand empathy and how emotions are perceived and communicated. I will first introduce the brain mechanisms that scientists believe mediate empathic reactions and how they were discovered. Then I will explore the dynamics of the actor–audience relationship and the techniques actors employ to charm audiences with their emotions. Lastly, I will also talk about how the brain distinguishes between reality and fiction and what happens in the brain during moments of absorption into fiction, those instants when we are magically transported into the world of imaginary characters.[9]

## A mirror for our emotions

The Spanish neuroscientist Santiago Ramon y Cajal (1852–1934) wrote: 'Human brains, like desert palms, pollinate at a distance.'[10] It is fascinating that he should be the author of such an affirmation, because his work paved the way for the understanding of how neuronal connections are established. Thanks to a silver staining technique developed by the Italian scientist Camillo Golgi, Cajal demonstrated that the nervous system is not an uninterrupted bundle of neurons wound around itself, as was generally believed at that time, but rather was composed of neuronal cells as separate units coming into contact through their ramifications. And we definitely need those neuronal contacts in order to empathize.

A new and attractive framework for the understanding of empathy emerged with the discovery of 'mirror neurons', cells that have revolutionized how we regard our emotional connections with others.[11] The discovery was as important and sensational as it was serendipitous. Back in the 1980s in a laboratory in the Italian city of Parma, Giacomo Rizzolatti, Vittorio Gallese and colleagues were investigating which brain areas were involved in the execution of movements. They noticed that a group of neurons in a region of macaque monkeys' premotor cortex, called area F5, fired when the monkeys performed a simple action such as reaching for a bite or grabbing a peanut. But F5 neurons were activated only if the movement involved an interaction between the agent of the movement and an object, and not if the movement had no specific goal or intention. Simply moving the arm with no goal was not enough for the neurons to scream their involvement in the detection instruments.

To deepen their findings, in the mid 1990s, the researchers then implanted electrodes in the monkeys' brains to record activity from *individual* motor neurons in area F5 while they gave the monkeys different objects to grasp. Here is where they faced a huge surprise. The moment they picked an object to hand on to the monkeys, the electrodes signalled some neuronal activity. To the researchers' amazement, the recorded activity came from exactly the same neurons that would also fire when the monkeys picked the same object themselves. Basically, the neuronal activity of observing an action *mirrored* the activity of performing the same action.[12]

These results were extremely thrilling because until then scientists had thought that the area F5 was involved exclusively in motor functions. Instead, the newly discovered mirror neurons displayed motor *and* perception capacities. When the monkey watched an action, even though it did not move a muscle to reproduce it, its mirror motor-perceptive system was activated as if the monkey were executing what it saw. In other words, the brain simulated action.[13] After these exciting discoveries in monkeys, everybody asked: do humans also have mirror neurons?

Applying electrodes deep into the brain of a person in search of single neuronal activity is not a feasible procedure. What you can easily do in humans is to use less invasive techniques such as fMRI. fMRI does not detect the electrical activity of single neurons, but the blood flow in the whole brain, so fMRI data would reveal areas that are active during both the observation and the execution of actions and *might*, therefore, contain neurons with mirroring functions. That is why in humans you cautiously talk about 'mirror-neuron systems' rather than single mirror neurons.

One of the first studies of mirror neurons in humans asked participants to watch experimenters make finger movements and then imitate those same movements. The results identified two cortical areas with mirroring functions.[14] One, located more towards the front of the brain, includes the inferior frontal gyrus (IFG) (Fig. 13) and the adjacent ventral premotor cortex (PMC). Another, located further back, is the inferior parietal lobule (IPL), which can be considered the equivalent of the monkeys' area F5.

The IFG is located within Broca's area, which is the brain's main language area. This suggests that the mirror-neuron system

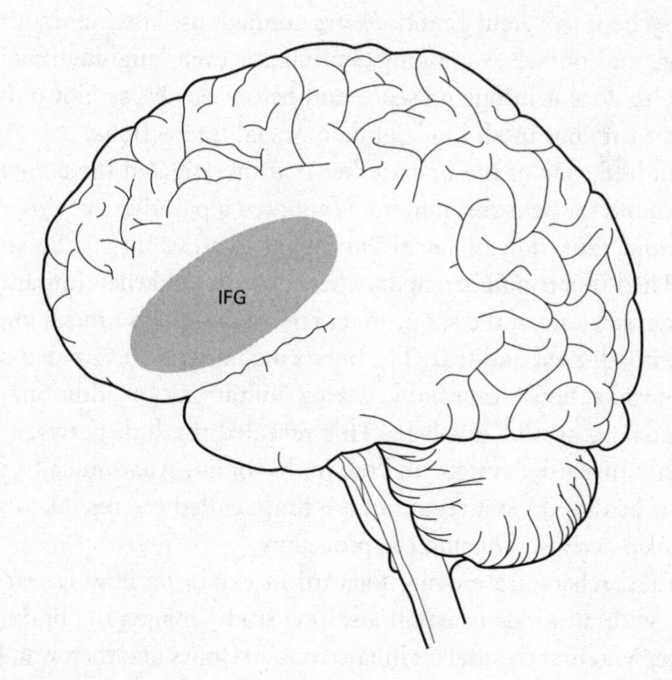


Fig. 13 Inferior frontal gyrus

may have been an evolutionary precursor of neural mechanisms for language. Speaking of evolution, it seems that the IFG may have evolved to be the common denominator underlying empathetic understanding across different emotions. A study testing which brain regions responded specifically to four basic emotions – happiness, anger, disgust and sadness – revealed that the degree of activation in the IFG correlated positively with the levels of empathy shown towards all of them.[15]

So, mirror neurons basically give us a second, more intuitive pair of eyes that shortcut the comprehension of the actions we witness. They allow us to apprehend an action we observe by making us simulate it in the brain. We *internally* know what someone else is doing.

This idea soon made researchers believe that the role of mirror neurons within the context of perceiving and simulating a simple action was only a tiny part of a more evolved mirroring system we use to empathize and understand each other's emotions. It only had to be uncovered! Emotions are contagious. How many times do we find ourselves cringing, smiling, or even laughing if someone else does it in our presence and before our eyes? Not only at the theatre, but in all kinds of daily social interactions.

Indeed, one of the first studies that investigated the empathic role of mirror neurons in humans adopted a paradigm of observation and imitation of facial emotional expressions.[16] The study consisted in letting participants first observe and then imitate facial expressions of the six primary emotions – joy, sadness, anger, surprise, disgust and fear. The mirroring network responded during both actions, especially during imitation. In addition, the amygdala was also involved. This revealed the link between the human mirroring system and the limbic brain. Anatomically, this link is achieved via a region in the brain called the *insula*, which was also activated during the procedure.

Researchers are moving forward in exploring how we empathize with all kinds of emotions. One study imaged the brains of people who first themselves inhaled nasty odours and then watched a film of an actor wrinkling up his face in a disgusted look. Both when feeling disgusted and when watching someone being disgusted, their insula fired up.[17] Even more interestingly, one study investigated 'tactile' empathy, that is, how we react to the sight of others being touched. Do we feel the touch ourselves? Indeed, the results indicated that the same area in the cortex would fire in people when they were lightly touched on their leg and when they watched video clips of others being touched in the same spot.[18] Most recently, another study revealed that the mirror-neuron system was involved in individuals who watched others yawn.[19]

## Mirroring the stage

The power of the mirror neurons has resonated widely within the theatre world especially, because it provides a fresh theory to probe the mysterious and tacit understanding between actors and audience.

The relationship between actors and audience is indeed theatre's raison d'être. When we view a performance, our whole bodies participate in the action before our eyes. We cringe in front of a horrifying act of violence or a display of disgusting material. Our guts contract during moments of suspense or fearful anticipation of danger. We get goose bumps at the sight of a moving act of heroism or a saddening scene of loss and separation. We almost feel the touch of a caress or a kiss being given on stage. Our skin and nerves relax when a conflict is resolved and harmony seems to prevail on stage. During a play, our mirror neurons are constantly at work.

The relationship between actors and an audience is osmotic. Both parties at either side of the footlights gain something. Actors spread emotions across the room. In turn, the audience provides a significant emotional feedback to the actors.

The eminent theatre director Peter Brook recounts a story that beautifully demonstrates the influence that different audiences can have on the quality and tenor of a performance.[20] His 1962 production of *King Lear* with the Royal Shakespeare Company was touring throughout Europe. Show after show, the quality of the production continued to get better, reaching a peak between Budapest and Moscow. Brook was endlessly charmed by how audiences with a scant knowledge of English could have such a profound and positive impact on the actors. At that time, the separation between the Western and Eastern sides of Europe was stark. Brook attributed the great response from audiences to an appreciation of the play itself, but also to a genuine desire for an interaction with foreigners.

These elements mingled together and manifested through silence and attentiveness that influenced the cast 'as though a brilliant light were turned on their work'.

The tour continued to the United States and the actors were charged with excitement and confidence that they would be able to offer an English-speaking audience all they had just learnt from the tour in Europe. When Brook attended a performance in Philadelphia, he was taken aback. The acting had lost most of the quality it had gained during the previous stops of the tour. The connection with the audience had drastically altered. Although the American audience could understand English perfectly, they did not engage with the play nearly as vividly as the Europeans. People yawned. Their motivation was different. For them, the production was yet another permutation of *King Lear* which they probably attended out of habit. As an audience, they needed something else. The actors did not ignore the requests of the new audience and responded by introducing a new rhythm. They highlighted every bit of moving or dramatic action by playing it in a louder and more pronounced manner. Ironically, all those complex passages that the Europeans had dwelt on and that an English-speaking audience could have easily absorbed were paced swiftly.

The audience is alive. While on one hand it can distract the actors (with noise, or with laughter at unexpected moments), on the other its silence and concentration or its synchronous response to a given moment in a scene may enhance and enthuse the acting.

The quality of such osmotic dialogue also depends on the type of theatre in which the play is acted. In most of today's theatres, in which the actors are blinded by the footlights, they can't exactly see what the audience are doing, nor look them clearly in the face. Most plays take place in dark theatres to create an atmosphere and to solicit the audience's imagination. Ben reminds me that this wasn't always the case and that plays have been staged in the dark for only about the last two hundred years.[21] In Elizabethan theatre, for instance, the connection between actors and audience was quite different in that the actors were much closer to the audience and actors and audience were equally lit. In the modern re-creation of the Globe Theatre actors can see the faces of audience members. They can play watching one person right in the eyes. This way, they can notice their reactions to the play, their amusement, sadness or joy.

Whether such direct exposure to the audience's reactions helps the actor may depend on the actor's skills and experience. Peter Brook called the audience a 'partner that must be forgotten and still constantly kept in mind'.

Yet what gives a performance emotional power, and how do actors work to create it?

## What's the trick?

In 1895, the playwright George Bernard Shaw witnessed something remarkable at a theatre performance. He was in London watching a play entitled *Magda* (originally *Heimat*, homeland), after the name of the protagonist, who on that evening was played by the gifted Italian actress Eleonora Duse.

Magda is a bold young woman who defies her father and escapes the bourgeois reality of her native town to venture on a career as an opera singer. While she is away from home, she enters a relationship with a fellow opera student who soon after leaves her alone with a child to raise (Duse actually experienced something similar in her own life). Having become a leading opera singer and a single mother, Magda decides to return to her native town and, overcome by a bout of homesickness, makes an approach to her father who agrees to take her back. A bewildering surprise awaits her at her childhood home. Soon after her return, she discovers that one of her family's intimate friends is the father of her child! In the third act of the play, Magda is on stage when her ex-lover is announced as a visitor. At first, she seems to cope well with seeing him again. They sit down and address each other cordially. But then, as Shaw noted in his review of the play: Duse (as Magda) visibly 'began to blush; . . . the blush was slowly spreading and deepening until, after a few vain efforts to avert her face or to obstruct his view of it without seeming to do so, she gave up and hid the blush in her hands'.[22]

Eleonora Duse was so intensely into character that she blushed on demand, as the theatrical moment required. Her performance made a huge impression on Shaw who was amazed by her ability to express embarrassment and discomfort so powerfully: 'I could detect no trick in it: it seemed to me a perfectly genuine effect of the dramatic imagination [ . . . ] and I must confess to an intense professional curiosity as to whether it always comes spontaneously.'[23]

Undoubtedly, Eleonora Duse was a fountain of talent. She was a theatrical sensation, both in her home country Italy and abroad. She had a unique gift for dramatic interpretation and, apparently, the rare theatrical authenticity that left G. B. Shaw so astonished came entirely naturally to her.

The question is: can the craft of acting out the details of an emotion with total authenticity be learnt, or even taught? Around the time of Duse's memorable performance in London, as she continued to command theatre stages in Europe and the United States, an ambitious and talented young Russian actor and director made plans to open an acting school. His real name was Constantin Alexeyev, but he was better known by the stage name of Constantin Stanislavsky. In 1897, at the age of thirty-two, he founded the historic Moscow Art Theatre, an establishment that was to become the cradle of a revolutionary method of acting.

Stanislavsky opened his theatre at a time of exciting and shifting views in science. The end of the nineteenth century witnessed the rise of psychology as a science – William James's theories, for instance, that I described in chapter 3. It's not clear what science publications Stanislavsky actually read himself, but his thoughts on acting shifted and evolved through the years, and were influenced by the science of his time. Eventually, he summarized his ideas in two remarkable books that are still the best place to go to grasp his enormous contribution to theatre, and are a pleasure to read.

An actor's emotions stimulate the audience. Stanislavsky described this as the 'irresistibility, contagiousness, and power of direct communion by means of invisible radiations of the human will and feelings . . . '. He compares those radiations to what is used to hypnotize people or tame wild animals. Similarly, he said, actors 'fill whole auditoriums with the invisible radiations of their emotions'.[24]

In his book *An Actor Prepares* Stanislavsky explains his acting technique through the story of a theatre director and acting teacher called Torstov and his pupils. The text is structured as a series of episodes, each constituting an acting class at his school in Moscow.[25]

One day when Torstov arrived at the theatre, he found the entire class intent on searching for a purse. He let them carry on with their hunt and watched them until they found it. Then he challenged them to repeat the search, so the students replaced the purse where it had been and started again. But the second time round the action was not convincing. It had none of the attentiveness and diligence Torstov had witnessed when the students were genuinely looking for something. The students protested, saying that their second search could have not been as effective because they knew exactly where the purse was, but Torstov insisted that since they were actors they should have been able to be just as convincing.

'We should [first] prepare, rehearse, live the scene . . . ' they objected.

'Live it?' Torstov said. 'But you just did live it!'[26]

During the early phase of his long theatre career, Stanislavsky had insisted on one crucial tenet: that he and his pupils fully embody a role. In Russian, this imperative was laconically summarized as 'переживание' (*perezhivanie*), or 'experience through'. Every actor must *become* the character he or she has been assigned. To achieve this transformation, an actor had to live through a part 'inwardly' and feel the emotions and sensations of the characters to be portrayed.

A central technique in the Moscow Theatre School was 'emotional memory'. Stanislavsky knew very well that while tiny details of events that occur in our lives can escape us, the emotions attached to them usually won't. Fear, dismay, hope, happiness, guilt can all be recalled, one by one.[27]

The Russian director invited his pupils to bring back to the surface memories of personal experiences and use them to portray the emotion of a character, just like a painter who, he said, can 'paint portraits of people they have seen but who are no longer alive'.[28] For instance, if they needed to express grief, they would do so by recalling the intense feelings of separation they had when they lost a close friend.

Stanislavsky didn't expect the recalled emotions to be identical to those experienced in the past. Though he demanded that actors be as sincere as possible he knew that the emotion recalled on stage was only a repetition. Emotions are fleeting and they flash by 'like a meteor'.[29] To colour their performances, his students were urged to draw from all kinds of sources beyond their own memories, such as books or travel, art, museums and conversations with other people. Even science. 'A suggestion, a thought, a familiar object' that bore personal relevance would help them revive the feeling.[30] But not just anything. Actors were invited to employ their imagination to select the most artistically powerful memories, those that were more 'enticing' and shared the highest affinity with the character. So, with the use of their imagination, and by drawing on their personal experience, actors needed to fully place themselves in the circumstances of the character they played. That their memories were distant in the past was not a disadvantage. Time, said Stanislavsky, was 'a great artist'. It could turn 'memories into poetry'.[31]

At a later point in his career, Stanislavsky felt something was missing in his practice as an actor and theatre instructor. He felt that his passionate teachings on how to incarnate a role on the basis of emotional memory and imagination demanded an additional framework. Evoking emotions from the meanders of an actor's past alone proved not to be a reliable strategy. He understood that when the actors spent too long carving out a character from the inside, they exhausted themselves and equally neglected sharpening the physical component of the performance.[32]

Stanislavsky needed a new source of inspiration. To find it, he resorted to science. Influenced by nineteenth- and twentieth-century reflexologists, he turned to conditioning theories to empower his methodology of acting. Stanislavsky sought a way to consciously trigger an actor's emotional expression through targeted physical cues. It was common knowledge that nervous pathways underlay complex behaviour and emotions, and that behaviour can be conditioned in response to a changing environment – let us not forget that Pavlov's ideas I described in chapter 3 were prominent in Russia at that time.

In a way, Stanislavsky became a scientist on stage. He realized that by selecting and carefully preparing key units of physical action pertinent to the logic of the character and the circumstances of the play, the actor could learn, by reflex, how to express the full-blown psychological experience of the emotion. In other words, physical action was the bait for emotion and the bridge between the actor and the role. He would ask his actors to proceed by accomplishing a sequence of small truths.

'When a whole action is too large to handle, break it up,' he said. 'If one detail is not sufficient to convince you of the truth of what you are doing, add others to it, until you have achieved the greater sphere of action which does convince you.'[33]

So, an action would be dismembered into its smallest physical parts and each part executed as truthfully as possible. A particular posture or movement would trigger a particular target emotion. Thus by working on minor actions such as clenching the fists and tensing the muscles of the neck, the actors would trigger anger, or they would produce feelings of despair by shuffling or sagging the shoulders. The body became the primary vehicle for the delivery of emotion.

Stanislavsky demanded something else of his actor pupils. Though they knew that the action on stage was fiction and not as true as in real life, they had to nurture a strong belief in their actions and their motives in order for those actions to be deemed persuasive by an audience. 'Truth,' he said, 'cannot be separated from belief, nor belief from truth.'[34] Everything that happens on the stage must be convincing to the actor himself. If it is not convincing for the actor, it will not be emotionally charged for the audience. For Stanislavsky, overplaying the 'truth for its own sake . . . is the worst of lies'.[35] To achieve his aim, he basically asked: what would an actor do *if* they were in the character's situation? An actor knows well he is not Hamlet, but what would he do if he were Hamlet? The 'if' worked to place the actors in the circumstances of the character, via their own imagination.

An actor's skills must be honed through practice. All units of action had to be rehearsed and practised for them to be reliably stored in an actor's baggage of experience – as any conditioning technique would require. Through repetition, the body learns to reproduce the emotion. As opposed to the internal search for the psychology of the character, the bodily experience constituted something more concrete, or easier to let recur, with the power to generate the real, full-blown experience, perhaps even capable of triggering a release of adrenalin to produce an on-stage blush like Eleonora Duse's. This was Stanislavsky's way to reach 'unconscious creativeness through conscious technique'.[36]

## The paradox of acting

Daniel Day-Lewis is renowned for taking his preparatory period for a role to some extremes. One of the remarkable components of his personal way of getting ready for a role is to refuse to break character during the production of his movies. He totally immerses himself in the life of the character he has chosen to portray. Apparently, he trained with a boxing champion for *The Boxer*, took butchering lessons for *Gangs of New York*. When he played a man with cerebral palsy in *My Left Foot*, he may have spent the entire filming time in a wheelchair and he successfully taught himself how to change a record with his toes. For *In the Name of the Father* he spent time in prison. When he played

Abraham Lincoln, an absolutely stunning, touching and convincing performance, he is said to have spoken in the accent and voice he created for the role even in between takes.

Such intense preparation is not a mere whimsical, eccentric way to get into the part. In an interview in which he discussed his method, Day-Lewis said that for his close adhesion to the character he needs 'to create a particular environment . . . the right kind of silence or light or noise. Whatever is necessary – and it is always different . . . '[37]

If you interpret that in Stanislavsky's terms, it is the construction of the right external physical conditions, that help support the overall experience of the role.

Duse's performance, and that of other acting geniuses such as Daniel Day-Lewis, is heralded as the epitome of *true*, *believable* and *authentic* acting.

Concepts of truth, credibility or authenticity are dangerous traps in theatre and acting in general. We expect a performance to be as convincing as something real, yet we know it is not. Actors know that, too. Any actress interpreting Medea is not going to actually murder her two children, nor is the palace of Corinth going to catch fire and burn down. Yet we are shaken by Medea's hatred and need for revenge. We fear her, and we also share her feeling of having been betrayed. An actor may be fully captured by Hamlet's vengeful rage, simulate the escalation of violence towards his uncle Claudius, but he will not ultimately nurture the actual desire to kill his colleague playing his uncle. Yet we feel the tension of Hamlet's hatred, we witness the run-up of his revenge. How can something be real and false at the same time?

As early as the eighteenth century the French philosopher and dramatist Denis Diderot recognized this paradox. In *The Paradox of Acting* he writes that an actor's performance of an emotion is not always the same thing as the feeling of the emotion perceived by the audience.[38] In order to be real, the actor must be artificial. In other words, in order to express an emotion and grip an audience with it, the actor must feel none. For Diderot, an actor must behave like an 'unmoved and disinterested' observer. He distinguishes between two main types of actors. One relies on what he called *sensibility*; the other on *intelligence*. Diderot's idea of sensibility is to play from the heart. But that kind of playing, he insists, brings no coherence. The playing will alternately be 'strong and feeble, fiery and cold, dull and sublime'.[39]

By contrast, the actor who plays 'from thought' and from careful study of human nature, will be one and the same at each performance and will always be at his best. The intelligent actor will have 'considered, combined, learnt and arranged' the whole play in his head. His 'passion' will have a definite course, with bursts and reactions, a 'beginning, a middle, and an end'. The 'accents' and the 'movements' during his performance will be the same.[40]

'What, then, is a great actor?' Diderot asks. 'A man who, having learnt the words set down for him by the author, fools you thoroughly, whether in tragedy or comedy.'[41]

Over a century before Stanislavsky, Diderot had outlined the challenges that the Russian would encounter along his own path, and had recognized the fact that an impassioned, internal search for the character would be prone to imperfections and that a more controlled, 'scientific' approach to acting would prove more reliable.[42]

In theatre, when we think actors are conveying emotions naturally they are actually conveying those emotions in the most unnatural way. When we believe they are showing us moments of great truth and utmost authenticity, they are outstandingly pretending. They are creating moments of great deceiving fiction.

On stage, truth and falsehood occur simultaneously, each a disguise of the other.

Stanislavsky said: 'A sense of truth contains within itself a sense of what is untrue as well.'[43] Whether truth or falsehood predominates in the scene depends on the actor's skills. We may be moved to tears by Romeo's pain and anger at Mercutio's death, but while the actor is clearly able to show the biological components of those emotions – he turns pale, he shouts – he doesn't always *feel* the emotions he conveys his character to be feeling.

Diderot uses a fine example that illustrates the core of the subtle distinction between reality and fiction. What is the difference between tears provoked by a real-life event and those evoked by a 'touching narrative'? – a question that even Hamlet asked after hearing the player delivering Hecuba's speech.

In response to a fine piece of acting, 'your thoughts are involved, your heart is touched, and your tears flow'. In response to a real-life tragedy, 'the thing, the feeling and the effect, are all one; your heart is reached at once, you utter a cry, your head swims, and the tears flow'. In the case of a real-life event, tears brim in your eyes suddenly, in the case of an acted one they come 'by degrees'.[44]

The magic of authentic acting is perhaps to reduce the distance between these two apparently opposed ways of feeling. As long as the desired effect is achieved, it does not matter what method is used. There may be both intense, inward characterization and a high degree of detailed groundwork at the same time.[45]

Daniel Day-Lewis says: 'I recognize all the practical work that needs to be done, the dirty work, which I love: the work in the soil, the rooting around in the hope that you might find a gem. But I need to believe that there is a cohesive mystery that ties all these things together, and I try not to separate them.'[46] Something about it will always remain mysterious.

## Reality or fiction?

On a daily basis, from childhood, we are constantly exposed to fictional worlds. We encounter fiction when we are told fairy-tales, when we read a book, when we play computer games or watch TV advertisements. And when we go to the theatre. The brain takes no break. It is extremely busy processing and integrating all this information, but it seems to have developed a way to distinguish what is real and what is unreal or fictional.

Dr Anna Abraham from the Justus Liebig University of Giessen in Germany was for a long time curious to map the neural networks that accomplish this task. She wanted to find out whether the brain operates by different mechanisms when it is exposed to a situation that is real as opposed to one that is entirely fictional. So she designed an interesting fMRI-based experiment that explored the brain's reactions to situations that involved either real or fictional characters.[47]

Participants were shown one-sentence written scenarios in which a real person named Peter was involved in situations that included George Bush or Cinderella. In one set of situations, Peter simply received information about both characters. For instance: Peter heard about Bush or Cinderella on the radio or read about them in the newspaper. The other set of situations involved direct interactions with the characters: Peter either spoke or sat down for a meal with them. What participants had to do was simple. They had to decide whether the scenarios portrayed were possible or not – that is, if they could indeed happen in the physical reality of the world we live in.

Obviously, it would be perfectly possible for Peter to hear about either of the two on the radio, but whereas Peter might actually meet George Bush in person, it would not be plausible for him to have lunch with Cinderella – at least not for real.

How does the brain operate when assessing these two different types of scenarios? The results were intriguing. Common to both types of situation was some level of mental activity in parts of the brain, such as the hippocampus, that are at work when we in general recall facts or events. Such activity was detectable regardless of the nature of the scenario – that is, whether the scenario was informative (when Peter only heard about the characters) or interactive (when he actually met the characters). However, there were a few striking finer distinctions in activity relative to the two scenarios and these depended on the type of character involved.

When exposed to scenarios featuring George Bush – a famous real person – the brain involved the anterior medial prefrontal cortex (amPFC) and the precuneus and posterior cingulate cortex (PCC). As I explained in chapter 1, the PFC is a wondrous region in the brain with multiple functions – such as keeping an eye on the limbic system, aiding our short-term memory and our attention. The amPFC and the PCC are medial parts of the brain that are involved in autobiographical memory retrieval, as well as self-referential thinking.

When fictional characters were featured, the brain responded somewhat differently. Parts of the lateral frontal lobe, such as the inferior frontal gyrus (IFG), were more active. The IFG is thought to provide mirroring capacities, but is also involved in high-level language processing. The fact that George Bush was linked to personal memory retrieval but Cinderella was not led the researchers to think that a crucial difference when assessing real or fictional scenarios might lie not so much in the degree of *realness* of the character involved, but in their *relevance* to our reality. To test this hypothesis, they peered into the brain of nineteen new volunteers who, as in the previous study, were asked to assess the possibility that a real protagonist could either imagine, hear or dream about or actually interact with a set of characters.[48] However, this time the characters involved in the scenarios were ranked in three categories with differing degrees of personal relevance for the participants: their friends or family (high personal relevance), famous people (medium relevance) and fictional characters (low personal relevance). As predicted, the activation in the PFC and PCC was indeed proportionally modulated by the degree of relevance of the characters described. It was highest in the case of friends and family members and lowest in the case of fictional characters.

The researchers gave the following explanation. When you encounter a real character, even if you have never met him or her, they will integrate into a wide, comprehensive and intricately connected structure in the conceptual storage of your mind. You are familiar with their basic behavioural features as human beings. You know more or less how they think, what kind of opinions they may produce. You are aware of the range of emotions that you can expect from them. By contrast, your mind is not equally familiar with fictional characters. No matter how much we know about the world of a fictional character there will still be something alien and inscrutable to us about that world. Take Harry Potter, for instance. You may have read all the books, but the amount of information you have gathered about Harry Potter – the hierarchy of wizards and the Hogwarts School of Witchcraft and Wizardry – is still definitely limited compared with the wealth of information that is available to you about members of your family, friends, or famous real people who are part of your immediate and past experience. Basically, in order to understand a fictional character, you need to dig deeper into your imagination, because he or she is bound up to fewer nodes of reference in your network than are real, or relevant, people in your life. Such nodes of reference as exist for the fictional character are also different in quality.

The fact that encountering a fictional character engages frontal lobe areas linked to language processing, such as the IFG, has an additional meaning. These areas are not responsible for understanding syntax, but more complicated components of language, such as semantics – that is, the meaning of words and symbols as well as other finer aspects of language such as metaphors. The fact that these areas are selectively activated when we encounter a fictional character implies that we are busy deciphering a whole new world that might be described with words and signs that require more than the simple decoding of syntax.

Abraham and her colleagues suggest that her experiments question what we mean by the *reality* of a situation. Reality is not just about what is ostensibly real or fictional for you. We do tend to distinguish between what is objectively real and what is fictional, but the distinction is much more subjective. If you live in Scotland, particularly if you're near a loch, the Loch Ness monster will be in some way real for you.

Basically if something is relevant for you, it doesn't matter if it is objectively real or fictional, it will be real for you, in your mind.

## Suspension of disbelief

In theatre, the boundary between reality and fiction is porous.

Throughout a play, we constantly switch between two worlds. One is the physicality of the boards of the set and of the actors in flesh and bone. The other is the fictional world of the characters and their story. When in a theatre, we witness actors in their corporeal appearance. We perceive their presence on stage. We hear their voices. If we are sitting in the front row, we may even feel their breath blowing towards us – as well as be met by some of their flying sweat. Simultaneously, as a parallel reality superimposed on that of the stage, we perceive and imagine the story being told. The set transmutes into anything from the palace of Thebes or the court of Elsinore to a cherry orchard, a battleground or someone's living room. We meet all sorts of different characters and we are introduced to their world. Some are well-known historical figures whose vicissitudes are deeply imprinted in our cultural background. Some are made up. Among these, some are more realistic than others, or rather, closer or more relevant to our own world.

Hamlet is a prince in Denmark. There may have been a Danish prince named Hamlet, but the one in the play is based on a legend and belongs anyway to a different historical time. Yet we understand Hamlet's plight. In Michael Frayn's play *Copenhagen*, on the other hand, we see on stage a theatrical representation of Nils Bohr and Werner Heisenberg, two great physicists who really existed. In *Death of a Salesman*, we face the struggling, desperate soul of a middle-aged man whose whole existence suffers a huge blow in one day, whereas other plays may shift across far wider timespans. Whatever the case, we always need to follow the story and temporarily adhere to the world of the characters, relate to them.

Theatre, and fictional representation in general, has for a long time employed a technique to reduce the distance between spectators and the characters: creating the circumstances in which spectators *suspend disbelief.*

The suspension of disbelief is a phrase first coined in 1817 by Samuel Taylor Coleridge (1772–1834). In his romantic poetry, Coleridge employed fantastic and supernatural characters that a rational, educated readership would not easily have identified with. Wanting to keep fantastic elements in his writing, Coleridge thought that by imbuing his narrative with enough facts and contemporary references he would help readers accept the story, rather than condemn it as implausible. He asked of his readers that they recognize 'a human interest and a semblance of truth' to the characters. He demanded 'a willing suspension of disbelief'.[49]

Unless you still believe in wizards, when you are enjoying J. K. Rowling's Harry Potter books you are also suspending disbelief, big time. In the specific case of theatre, suspension of disbelief is achieved by believing that in addition to the three walls of the set there is a fourth transparent wall separating the audience from the action on stage. Erecting such a wall secludes the play into an independent box. The actors go on with their scenes as if nobody were watching them and the audience believes the world of the characters is real, despite it being played on a stage.

In the prologue to *Henry V*, Shakespeare begs the audience to forgive the bare stage and use their imagination to picture it as the world of the king in war with France:

> But pardon, gentles all,
> the flat unraised spirits that have dared
> on this unworthy scaffold to bring forth
> so great an object: can this cockpit hold
> the vasty fields of France?
> . . .
> Piece out our imperfections with your thoughts;
> . . .
> and make imaginary puissance . . .

Give power to the imagination to accept, and then ignore, the illusion of a reality that is not real.[50]

The suspension of disbelief is not a universal aim in theatre. The great twentieth-century German playwright Bertolt Brecht (1898–1956) deliberately turned this tactic on its head and had specific expectations of the relationship between actors and their audiences. Brecht believed theatre should not force empathy. He disliked audiences that would passively absorb and believe in the events of the story embodied on stage. He was deeply frustrated with the majority of the traditional theatre of his time. Provocatively, he used to say that traditional theatre turned the audience 'into a cowed, credulous, hypnotized mass'. He even said that the 'audience hangs its brains up in the cloakroom along with its coat'.[51]

On the contrary, he ensured that his audiences became occasionally and strategically detached from the scene. Bertolt Brecht introduced the theatre technique of alienation – originally *Verfremdungseffekt* in German. He wanted his audiences to breach the fourth wall and become aware they were witnessing fiction, not a real-life event.

As I mentioned at the beginning of the chapter, theatre is a tremendously powerful vehicle for portraying the world we live in. It can be used to denounce the problems afflicting our society, sometimes in a satirical fashion. The ultimate aim behind Brecht's revolutionary staging choice was to empower the viewers to critically question the social realities represented in the play and see them in a new light. He encouraged dissent from the action and the freedom to judge it. For instance, Brecht's *Mother Courage and Her Children*, written in response to Hitler's invasion of Poland, is a play set during the Thirty Years' War that condemns the rise of Fascism and Nazism.

Some of the elements Brecht used to interrupt the flow of the play were simple. For instance, by having an actor stand beside a bare placard on a stage deprived of even the most basic items of scenery, he reminded the audience that they were in a theatre. He also had actors sing out of character or address the audience directly by introducing pieces of text that were not part of the main body of the story. Occasionally, he would bring the lights up in the auditorium. In all cases, the adhesion to the story would cease temporarily and the audience was invited to abandon their dreamful state and judge the social reality of the characters portrayed. In Brecht's plays, the characters are not always or entirely who they are supposed to be. In other words, the actors disconnect from the part they are playing.

Although the alienation effect disrupted the flow of the

drama, Brecht was not aiming for an absence of emotional transfer. If in traditionally realistic theatre the transfer of emotion is achieved through the superimposition of actor and character, in Brecht emotion stems from their divergence.

During the last hundred years theatre has certainly escaped from the confines imposed by traditional dramaturgical rules – such as causal linearity, plot, plausible characters – both in the writing of theatre texts and in staging choices. Fragmentary scenes made up of sounds, images, movements and games of lighting that do not require the adhesion to a realistic story can be laden with poetic metaphor and symbols of equal emotional power. Even the selection of the physical space for a show has become significant, the traditional four walls often being abandoned. A story may be told in a small intimate room, in big arenas, across multiple rooms, or played in spaces that lend themselves as excellent metaphors for the meaning or content of the play. Emotions flow in a theatre room not solely through a story being told in words and acted from start to end.

For three entire months, from March to May 2010, the acclaimed artist Marina Abramovic, the 'grandmother of performance art', sat for seven and a half hours every day on a chair at the centre of a large room in the Museum of Modern Art in New York. The work was entitled *The Artist Is Present* and was the central piece of Abramovic's retrospective at the museum. In front of her was another chair on which, one by one, visitors sat to face her gaze. Each encounter was unique, but followed a simple ritual: when a visitor stood up from the chair, Marina would close her eyes and slightly duck her head into her breasts while waiting for the arrival of the next guest. Then, as soon as the latter took his or her seat on the chair, she would slowly raise her head and look right into their eyes. In the course of three months she looked into 1,565 pairs of eyes. Many wondered: what was going on there? What was her aim? And: was that theatre?

In an interview following her show Abramovic settled the issue by firmly expressing her distaste for theatre because of its fakery: 'to be a performance artist, you have to hate theatre.

Theatre is fake: there is a black box, you pay for a ticket, and you sit in the dark and see somebody playing somebody else's life. The knife is not real, the blood is not real, and the emotions are not real. Performance is just the opposite: the knife is real, the blood is real, and the emotions are real. It's a very different concept. It's about true reality.'[52]

Abramovic speaks from personal experience. She is renowned for having employed real knives to pierce her skin in front of an audience, in addition to having put herself at risk of death in some of her performance pieces.

But theatre is not just fake. Theatre is fake and real at the same time. So is Abramovic's performance, we could say. We know that the person who is sitting on a chair at the MoMA the whole day is the artist Marina Abramovic, but during her performance our imagination does not shut off. That person could be just a character, a mysterious and charismatic woman with beautiful long hair and a long red gown who has lost her speech. That a woman would choose to sit in a room for three months is plausible, but also not very common. The encounter between Marina and her guests was a frontal exchange, in which spectators had a chance to constantly switch between two planes of reality. That normally happens in theatre, too. And, as I explained above, as human beings we are equipped to distinguish between real and false, the actual and the imagined.

But whether or not you call Abramovic's courageous and elegant performance theatre, empathy was certainly at work across those two chairs. Most of the people who confronted Marina's gaze became emotional. Many of them shed tears. A few sobbed. It's important to remark that Marina gazed her visitors directly in the eyes. Actors on stage rarely have the chance to gaze into the eyes of their audience. This has an interesting scientific implication. The facial broadcast of our emotions is a fundamental vehicle for conversations between minds. But the brain reacts in sharply different ways when we simply glance at someone's face and when we look right into their eyes. The eye gaze, after initial processing in subcortical regions of the brain, goes on to stimulate structures that modulate our social interactions.[53] In addition, only direct eye-to-eye contact activates areas such as the dopaminergic system that induce reward and inspire proximity.[54]

In sum, whatever nature of performance you watch, there will always be an emotional filter to it, and a veil of illusion.

## Carried away

Illusion is a crucial element throughout a performance.

One study specifically looked at the nature of illusory moments in theatre, those instants in which we forget where we are. Researcher and theatre director Yannick Bressan and collaborators in France explored the blending of reality into fiction in the context of theatre with a creative fMRI experiment, in which participants watched a live performance while their brain was being scanned and their heart rate measured. The goal of the study was to discover what brain regions are active during moments of adhesion to fiction.

The live performance was a monologue, adapted from a contemporary dramatic poem, *Dionysus the Wild*.[55] The mythological being of the work's title, half man and half divinity, is the protector of the grape harvest and winemaking and the guardian of mankind's basic instincts, associated with madness and excess – certainly a figure well acquainted with passions and emotions. But he is, of course, a myth, a fictional character. In the play, he is bizarrely stranded on a New York subway platform in the year 2000 and tells his tumultuous life story, recounting epic travels through cities of ancient times.

The researchers and the theatre team wanted to make the act as close as possible to a real theatre performance and to re-create an environment in which the viewers would feel engaged in the story from start to finish. So, as each of the participants was being prepared to enter the scanner, an actor in the room would begin to recite the monologue. When the scanner bed was made to slide into the magnet, the actor went to act in an adjacent room, but the viewer inside the scanner continued to watch him through prismatic goggles connected to a screen where the scene was being played. When on, a brain scanner emits loud, disturbing noises. To avoid distraction and interference with the appreciation and understanding of the monologue, the researchers and the theatre team cleverly incorporated the noise of the scanner into the performance, by staging it as drones of trains reaching the subway platform where the fiction was supposed to take place.

How did the experiment identify moments of adhesion to the performance?

Prior to the experiment, the theatre director had selected twenty-four 'events' within the written text that were intended to elicit a shift in the viewer's perception of reality, from the actual physical reality (that of the scanner and the experiment room) to the fictional reality of the monologue. These elements worked as adhesion-to-fiction 'markers' throughout the play and were highlighted to the actor and the production team as a list of direction instructions that included movements, voice tones and intonations, sound, lighting and other kinds of scenery effects.

A few of these markers corresponded to salient passages in the story of the god's life that alternated moments of fierce rage and calmer moods, all told and staged very dramatically. For instance, at one point Dionysus recounts his own death. The rhythm of his speech is faster and the tone of his voice more solemn. Later, Dionysus comes back to life. His rebirth is symbolized by the appearance of a light in his hands, which he protects like a precious object. Charged with rage and driven by fury, he takes revenge by killing the men who slaughtered him. During these moments, Dionysus behaves more like a beast and moves quickly, speaks loudly, stares aggressively.

At the end of the scanning procedure, participants were invited to report their subjective experience of the performance while they watched a recording of it made when they were inside the scanner. They were asked to describe their thoughts and feelings about the monologue. Their comments were annotated for every five-second period of the play. After commenting on the whole play, they were also asked questions exploring their involvement in the piece, some of which were specifically intended to probe their adhesion to fiction, e.g. whether or not, and at which specific moment, they were able to disregard the experimental set-up, when they literally felt transported into another reality, or if and when they believed during the play that they were in the presence of Dionysus and not the actor.

This in-depth subjective reporting permitted the identification of moments during the play in which the viewers felt transported into another reality. Since fMRI and heart-rate data were acquired throughout the duration of the play for each of the viewers in the study, it was possible to link any moment of adhesion to fiction to relevant changes in brain activity.[56] For the purposes of the experiment, moments of adhesion to fiction were defined as instances where the spectator's offline subjective report coincided with one of the stage director's selected 'marker' passages – that were intended to solicit the adhesion – the time-point within the performance matching exactly.

Remarkably, 69 per cent of the elements in the play subjectively experienced as adhesive to fiction coincided with the elements predefined by the director. Of these, 40 per cent were textual elements and 60 per cent consisted of more directorial markers, such as the use of lighting or the movements and expression of the actor.

The brain regions that fired in moments when fiction blended with reality were several. One was the inferior frontal gyrus (IFG), which comprises the mirror neurons, processes language and is involved in recognizing motion and in the interpretation of facial expression, things that are essential in theatre.[57] Another was the posterior superior temporal sulcus (pSTS) (Fig. 14).

Like the IFG, the pSTS plays a role in our ability to understand other people. Interestingly, when someone sustains damage to the pSTS it becomes difficult for them to accurately assess where another person is gazing or interpret what they are feeling about the object they are gazing at.[58] The pSTS also governs the comprehension of language, including text and verbal processing, specifically the ability to understand metaphor.[59] It would be surprising, therefore, if the pSTS were not active, since watching a

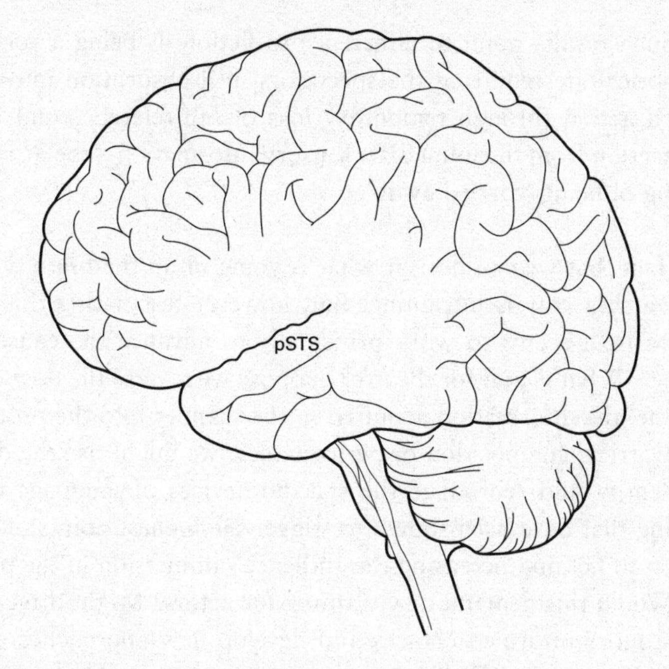


Fig. 14 Posterior superior temporal sulcus

play – one rich in text – involves a high degree of language comprehension and the appreciation of a poetic and metaphoric use of language. These same areas have also been shown to be involved in processes of social and aesthetic judgement.[60] Where theatre-watching is concerned, this function probably has a role in aesthetic appreciation of the writing style, the plot or the characters of the play, its overall staging and direction.

Concomitantly, during the adhesion moments there was also a decrease in heart rate and reduced activity in midline cortical areas such as the dorsomedial prefrontal cortex (dmPFC) and posterior cingulate cortex (PCC), which are normally engaged in representation of the self, also in relation to the external world (in chapter 2 I explained that in its function the dmPFC roughly corresponds to Freud's ego). Absence of activity in these areas would blur the boundary that distances us from the reality of the story being enacted. We are helped to get closer to the fiction.

Such results point to adherence to fiction as being a sort of hypnotic state requiring the spectators' full absorption into the staged action through temporary loss of self-reference and disconnection from the immediate sensory information – the distinct feeling of being 'carried away'.

The fact that we can peer at what is going on in the brain when we watch a play is intriguing. But, however fascinating this research is, it seems to work principally to advance the cause of science. What is in it for theatre? Suppose we reverse the flow and channel the information acquired in the scanner into the process of theatrical composition or performance: we might use the data to identify and reproduce the specific devices of language and staging that have been shown to trigger the highest points of adhesion to fiction, increasing the audience's immersion in the play.

Would this demand new training for actors? Might directors make more informed choices and develop new approaches that are audience-oriented? What types of movement or expression are most poignant when we try to convey grief, anger or joy? What metaphors work best to compress an action or thought? What elements of plot device, vocal emphasis or even lighting provoke an alteration in the spectator's brain activity?

While this might sound like an exciting, novel possibility, I remain sceptical that one has to dissect a theatre piece into units and put them to the test of neuroscience and brain imaging in order to ascertain their effectiveness.

So does my friend Ben: 'I don't know necessarily what it is that I do that would make an audience laugh or cry, but I know how to do it. It's a raw instinctive thing that I have been training over the years and that has been honed with skill and technique and craft. In some respects, I don't think I want to know, because I would be worried that it would become too technical.'

Anyone who has worked in theatre knows that fMRI images and good statistics could never fully substitute for the unpredictable and revelatory power of a rehearsal room.

Writing and acting a theatre scene, or deciding whether it

'works' or not, is for the most part a visceral process which, despite being based on technique, craft and experience, maintains a high degree of inexplicable subliminal intuition, which has proved successful for centuries. Theatre artists will continue to exercise their metaphors and explore infinite ways of playing with them as they have in the past. Knowledge of the mechanics of mirror neurons and other brain areas can only add so much to the ability of directors and actors. Perhaps only emotions can generate emotions.

## Coda

The lights go down abruptly. Darkness signals the end of the show. A few moments of hesitation, then everyone takes a breath before exploding into a loud choral applause. The lights come up again, blinding Ben's eyes.

The end of a performance is always a sad moment. Theatre is a ritual of death as much as it is one of birth. The concentration, the involvement in the action, the height of emotion and the intensity of the invisible communication between the audience and the cast across the footlights all gradually vanish. The magic evaporates. I don't like letting the characters go. I wonder how it must be for the cast to let go of them when a production ends.

There was really no moment during the play when I thought about my brain and what it was doing. When I am moved by an actor on stage, I know that his or her captivating performance is altering my cerebral activity, but the thought of such alteration will neither enhance nor weaken my emotional status.

But I do remember moments when my eyes thinned when I smiled, when I jumped in response to a shout, when my throat began to close when I saw grief. And I remember the moments when I forgot my surroundings.

Peter Brook condenses the magic of theatre into one sentence: 'In everyday life, "if" is an evasion, in the theatre "if" is the truth.'[61] It's really about being exalted, about dreaming, falling prey to an illusion. It is about living in constant evasion.

# 6

# Joy: Fragments of Bliss

Nothing is funnier than unhappiness, I grant you that

SAMUEL BECKETT

Count your age by friends, not years
Count your life by smiles, not tears

JOHN LENNON

Manhattan, five o'clock in the morning. After hours of burning the midnight oil, I finally put the pen down. This time not because I didn't know how to carry on, but because I was actually done with the writing. I wasn't abandoning the page with frustration, hoping for a better season. I was finally harvesting the crop.

A source of pleasure for me is to write a poem every now and then. I use verse to condense pieces of my life into short cherishable fragments, ornate strings of words I can easily look back to, repeat to myself and share with others to make sense of changes in the way I look at life. Occasionally, it is a strategy to dress a sore experience in a comfortable disguise – even mishaps assume beauty in poetic form. But in general, it's just a way to keep my passion for language alive and challenge my skill at transforming emotions into words, mental understanding into written discourse.

My favourite form of poetry is the sonnet and when I landed in New York, a city that infallibly puts me in a good mood, I was right in the middle of creating one. For a week, I had laboured over rendering into this old form of writing the evolution of feelings I entertained for someone. I wasn't at all sure where our mutual infatuation was leading, but I was sensing some kind of transition, an elevation of some sort: from an insecure ground to a plane of optimism. I could see the emergence of some confidence, the tip of something joyful, and I wanted to celebrate that.

I was determined to finish it, feeling I was close to something, but who can command the creative process? I had worked on the sonnet on the plane – I usually get good ideas when I fly – writing the lines across two pages in my notebook, marking the tonic syllables of each word boldly. I had composed the first eight lines, but the remaining parts of the sonnet were still a chaotic set of ideas that needed to find space to fit into this fixed structure, with rhyme and everything. Anyone attempting to create something knows well that moments of success alternate with moments of frustration. On the page, the broken lines looked like this:

Suspended in such spell, we . . . ?<br>
The truth descended from our yearning eyes<br>
? . . . Resisting afterthoughts, . . . ?

The final couplet was missing in its entirety, too. But I knew I could find a solution, if I hung on.

I raised my head from the notebook and paced up and down the room a couple of times. As I walked, I sensed some cracks on the wooden floor beneath my feet. Then I stood by the window and looked up to the sky; a strip of the Hudson was visible in the far distance. It had been misty all evening, but the wind had pushed the clouds away. The whole city was about to wake up. How I simply loved being in New York. As most lights in the buildings went off, I gazed at the last fading stars. And that's when ideas started to come back. The cracks on the floor. Cracks are like the scars of desire that I was longing to heal. And the stars . . . Of course. Stars rhymes with scars. I still didn't know exactly how, but I knew that was the road to take in order to fix the lines, and that it made sense to do so. Truly, never had the sky been so beautiful and full of promise. So, I set to finish, trying not to let the momentum fade away.

The missing piece in the puzzle finally surfaced. Scattered fragments united to form a continuous sentence without gaps. Chaos surrendered and made space for more order. Dissonance blossomed into a song and I even found the words for the concluding couplet. The poem was finished and it sounded well, at least well enough to me:

Suspended in such spell, we won high tides
  Embraced the water, gazed upon the stars
The truth descended from our yearning eyes
  Resisting afterthoughts, erasing scars
Here, tears are sweet, well then what gives to cry?
At sea, through the night, you and I fly high

Each time I conclude a piece of creative writing – any piece of writing, in fact – I can't believe what has just happened to me. I didn't have a mirror, but I bet my forehead was relaxed, and a sparkle of light must have tinged my eyes, coating them with pride. When, after an erratic wandering of the mind, the right word is on the page, a sentence takes shape before my eyes on paper, I feel a gush of joy. A blow of satisfaction. Perhaps, the joy originates in the clarity of mind. Excited as I was, how could I go to bed? Despite being tired, I was willing to celebrate the event, so I walked to the river, whistling all the way.

## Last but not least

We are finally dealing with enjoyable emotions. I have first covered negative emotions and left the positive ones to the end, because I naturally thought it would be best to challenge you at the beginning and then leave you with a sweet taste in your mouth rather than the other way around – *dulcis in fundo*, as the Romans would say. It is also true that, unfortunately, science hasn't dedicated as much attention to enjoyable emotions as it has to the negative ones. We know much more about anger, fear, disgust and sadness than we know about emotions that uplift us, such as joy. Fear is by far the emotion that has been studied most extensively. Research on joy and happiness really only started to be undertaken seriously in the 1990s. The reason for such discrepancy may simply lie in the aspiration to understand negative emotions so that we can best avoid or interfere with them.

At the beginning of the book, I briefly mentioned that, as biological creatures, we have two basic survival mechanisms at our disposal as we navigate through our emotional life: approach and avoidance. Such mechanisms are opposed strategies that have been shaped by years of evolutionary development and are shared by organisms as diverse in their complexity and sophistication as are an amoeba and a human being. The rules are pretty simple: pain is to be avoided, pleasure is to be pursued. These two fundamental tenets have been pillars of shifting scientific and philosophical theories for millennia. Even of psychoanalysis. Freud summarized this polarized view of emotional regulation when he pondered what men and women demanded of life: 'The answer to this can hardly be in doubt. They strive for happiness; they want to become happy and to remain so. This endeavour has two sides, a positive and a negative aim. It aims, on the one hand, at an absence of pain and displeasure, and, on the other, at the experiencing of strong feelings of pleasure.'[1]

It helps to regard ourselves as organisms who constantly seek to be in a fine equilibrium with the environment. We strive for a balance – which in scientific language is called *homeostasis* – and our actions and behaviour are movements that make us swing from one experience to the next in search of this equilibrium of well-being. Life is full of obstacles as well as reasons to be happy and we veer from one type of incident to the other. Some episodes are more painful than the average. We encounter the pain and we run away from it, towards a more pleasurable experience, but then we might incur pain again. Say we find shelter from the pouring rain under a tree. Everything seems great until a mosquito bites us. Or we wake up in a great mood, we run to the bakery to fetch an amazing fresh croissant, bump into a friend and then we sit down at the desk to start work only to find that our computer has crashed – this did really happen to me once. Viewed from this angle, pleasure is what we gain from a swift departure from pain as we approach the equilibrium again.

Indeed, pleasure can become painful and pain can occasionally give satisfaction. Sadistic sexual activities can provide joy to those who practise them. The sight of a delicious chocolate cake in a baker's window is pleasure, but if we ate an entire cake on our own, that same cake would probably be a source of discomfort. Love is a reason for joy as it is for sadness, especially when it ends, causing grief. Also, the intensity of pleasure and pain is always related to what state of pain or pleasure we are in already. If we are in deep pain, what would normally be a small pleasure can elevate us to an ecstatic state.

I am going to tell you about some of these peculiar aspects of pleasure, as well as some of the roads that can help us reach joy. But first, as I have done for all the other emotions dealt with so far, I want to tell you what joy looks and sounds like.

## Signs of joy

A smile gives joy away. Intuitively, one would think that the manifestation of a smile is accomplished around the mouth. Indeed, one of the muscles at work in a smile is the zygomaticus major, the muscle that extends from the cheekbones down to the corner of the lips. But the contraction of this muscle alone is not enough to produce a recognizable genuine smile. The first to report this was Duchenne, the French doctor who stimulated facial expressions by placing galvanizing electrodes on people's faces and whose pictures Darwin used to illustrate his book. What led Duchenne to identify the rest of the scaffolding behind a truthful smile was the telling of a joke. When the French doctor applied his electrical stimulus to the zygomaticus major alone, his subject's resulting facial expression looked unnatural and the smile false. When instead Duchenne told the man a joke, the amusement painted a totally credible smile on his face.[2] Guess where the difference lay. In the man's eyes. When a smile is sincerely joyous, the muscle around our eyes, called the orbicularis oculi, also contracts. What all this means is that while you can voluntarily thin and extend your lips to produce a smile that communicates politeness, for instance, you can't just move the orbicularis on demand. Hence, you can't fake a joyful smile. Only true enjoyment produces a complete smile, which is still referred to as a 'Duchenne smile'. Such fine distinction is reminiscent of the importance of the contraction of the muscles between the inner ends of your eyebrows, in addition to the down-turning of the lips, for a complete expression of sadness.

There are few things more embarrassing than finding yourself laughing uncontrollably when you really should be keeping a straight face. Unfortunately, it happens. Someone you are interviewing shakes your hand and introduces himself as Constant Pain. Your boss greets you after the lunch break unaware of the fact that a tiny piece of spinach is stuck right between his front teeth. Somebody falls clumsily in front of you.

Laughter is not only a sign of joy or amusement. Laughter can be cynical, malevolent, deriding. It may even accompany violent acts, such as killing.

But in any case, laughter is also more than just an open grinning face. When we laugh our lungs, larynx and the muscles in between our ribs are at work. So when examining laughter, we also explore the sound of emotion, in addition to its visual appearance. Laughter has a voice. And laughter, if listened to carefully, has a distinct acoustic signature. The psychologist Robert Provine has dissected the structural components of laughter.[3] To do that, he had to listen to a lot of laughing. It's not easy to have people laugh on command, but one of the strategies he adopted was to go around meeting people in public spaces, telling them he was studying laughter and asking them to laugh. The reaction to that statement was often a spontaneous laugh and he recorded those. When he unreeled the tape and analysed the sounds in a laboratory, using an instrument called a spectrograph, he noticed a distinct pattern. Laughs are made up of a series of vowel emissions – mostly ha or ho – which are repeated at evenly ordered intervals of time. The duration both of the laughing syllables and of the intervals is to be measured in milliseconds. Another interesting characteristic he could observe is that laughs are not scattered in disorderly fashion into our conversations. They often follow sentences, they don't interrupt them. They work like punctuation marks. In general, Provine believes that we must have developed distinct neural circuits that make us detect and process the structure of laughter and then generate it via the same type of vocalization, making the contagious aspect of laughter possible.

Besides being contagious, laughter is also universal. There are sounds of laughter across the animal kingdom. For instance, chimpanzees laugh, although the breathing pattern in their laughs is different from the pattern observed in humans. Even rats laugh, especially when they are young. Their laugh is obviously nothing like ours, neither are rats renowned for having a sharp sense of humour, but they do emit measurable ultrasound vocalizations in pleasurable circumstances. When 'adolescent' rats are at play with one another and when they are tickled on their back, neck or belly, they emit distinct ultrasound vocalizations, squeaky noises with a frequency – about 50kHz – which is higher than the frequency of vocalizations emitted in anticipation of aversive, unpleasant circumstances (about 20–30kHz).[4]

The cognitive neuroscientist Sophie Scott, of University College London, has for a long time been interested in understanding how we communicate with one another, both through the production and perception of speech and through other forms of non-verbal exchange. She and her team have generated beautiful data on laughter.

Two of her collaborators travelled far to find proof of the cross-cultural nature of emotions. This time, their interest wasn't in facial expressions, but the sounds of emotion. They reached a few remote, isolated villages in North Namibia, where the inhabitants, the Himbas, had never been exposed to cultures other than their own and, therefore, were not familiar with the emotional signals of Western Europeans.[5] Basically, they had never had the chance to hear a Londoner cry or laugh. The experiment went as follows. The Himbas listened to stories (in their language) that focused on a few basic emotions, such as fear, anger, sadness, disgust or amusement. Then for each story they heard two sounds produced by English speakers – one matched the story (and the emotion), the other did not – and they were asked to identify the right one. When the researchers returned to London, they brought back with them recordings from the Himbas and tested an English group of participants in the same way. Both the English and the Himbas group of listeners recognized the sounds connected to the emotions quite consistently. In the case of amusement, which was exemplified by a tickling scenario, both groups unequivocally matched laughter to it. The Brits detected and recognized the laughter of the Himbas and the Himbas recognized that of the Brits without fail, both associating laughter with tickling, which, as we know, is often a source of joy, even for rats. Laughter, then, is the acoustic equivalent of the smile. It is another marker for joy as a universal emotion.

Sophie Scott has also deepened her understanding of positive emotions by looking for the neural clues to the strong contagious aspect of laughter. In chapter 5, I talked about the power of mirror neurons to propagate emotions between actors and an audience and in general about the power of facial mimicry to imitate expressions. As one might expect, laughing in the presence of others entails incredible mirroring activity. But Sophie Scott and her collaborators showed that even the sound of laughter, and not just a visual stimulus for laughter, can engage mirroring parts of the brain and generate homologous facial expressions in the perceivers.[6] In fact, of the many emotional sounds she used to probe the auditory capacity of the mirroring system, laughter was the most powerful. Basically, just hearing someone laugh can prompt a smile on your face.

Finally, laughter is definitely a social expression of emotion rather than a solitary activity. We may occasionally laugh on our own in front of an amusing comedy, but laughter is mostly a social affair. When the psychologist Robert Provine asked a group of students to keep a regular diary of their laughing during a whole week, the results were clear. The entries for their laughs revealed that they laughed thirty times more in the presence of others than in solitude.[7] Laughing with others assumes all sorts of social meanings. We laugh to agree with others, to bond with them, to show them our trust and our love.

I must confess something. I enjoy a good laugh, especially with friends, but in me the true sign of joy is whistling. If I am in a good mood, or I want to get into one, I can whistle you a whole symphony.

## An entanglement of pleasure and intellect

Let's go back to my fleeting pleasurable moment of creation at the dawn of a New York City day. Writing a poem, composing a song and other kinds of intellectual and creative endeavour are indeed pleasurable activities. I did gain gratification from chiselling out my sonnet at five o'clock in the morning. But how did it happen that I came to make sense of random thoughts and images, and finally grasp what was missing in the poem?

Personally, so long as it keeps happening with sufficient regularity, I prefer to keep a good part of that creative process as an unyielding mystery. However, research is beginning to uncover some of the mechanisms behind such mental processes and the findings, even if perhaps preliminary, are fascinating. One main lesson emerging from laboratory data is that positive mood is coupled to the achievement of sharpness in the mind. Even just a short lift of mood improves our capacity to think and our creativity.

I will concentrate on this later, but for now let's take a step back and explore the basic anatomy of pleasure. The brain has a centre dedicated to pleasure that is habitually called the 'reward system'. Because of the ancient evolutionary purpose of sensory pleasure, the reward system is a primordial device that has been an essential part of the brain, and not just in humans – to give you an idea, bees, rats, dogs and elephants all have comparable reward systems. If in a bee the reward system consists of a single neuron, in higher animals it comprises several tissues.[8] In Fig. 15 I have highlighted the relevant tissues in the human brain: the ventral tegmental area (VTA) and nucleus accumbens (NA). The

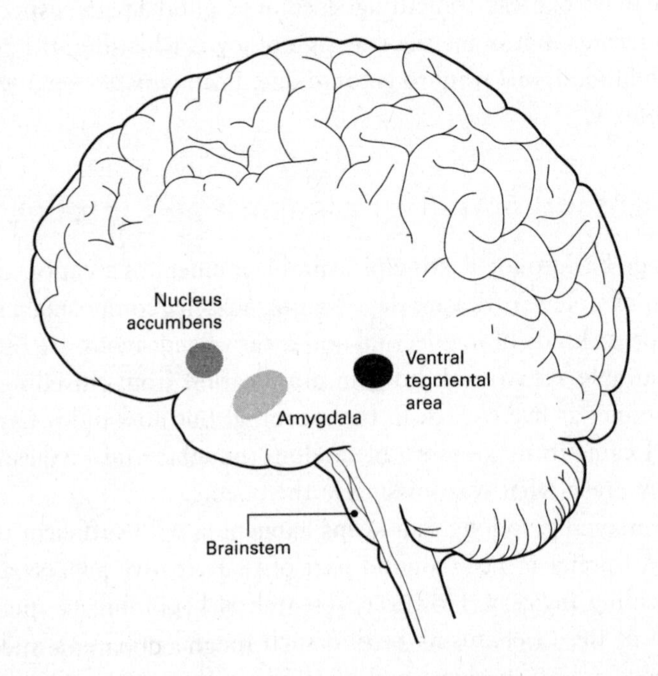


Fig. 15 The nucleus accumbens and the ventral tegmental area are part of the reward system

VTA is part of the brainstem that covers the top of the spinal cord – in Latin *tegmentum* means cover. The NA takes its name from the fact that it leans – *accumbens* – towards the septum, a smaller region of the brain just above it.

The proper functioning of the reward system makes sure that essential behaviours such as eating or sex are experienced as satisfying, and hence likely to be repeated, allowing survival and reproduction. Rewards coming from given stimuli and actions reinforce our desire to increase the frequency and intensity of those stimuli and actions.

The pleasure-inducing capacity of the reward centre was first observed in rats in the 1950s. The researchers gave the rats a mild electrical stimulus each time they moved to a given corner within their cage. The current was delivered through an electrode inserted in the septum area – that is, behind the rats' nose – that reached down to the reward areas. The stimulation turned out to be pleasurable for the rats because instead of avoiding it, they spontaneously returned to the same corner repeatedly.

Later, the researchers connected a lever to the source of the electrical current so that if the rats pressed it they would stimulate themselves. The rats were so avid to receive the stimulus that they couldn't stop pressing the lever. They would press it hundreds of times an hour – it's a conditioning process similar to the one I described in chapter 3, but this time the stimulus is pleasurable not painful, hence carrying a positive incentive.[9]

One of the ways the neuronal cells in the reward circuit communicate with each other is by sending and receiving a neurotransmitter called dopamine, which acts as a messenger in the same sort of way as serotonin – which I described in chapter 4.

Briefly: upon stimulation, dopamine is released from a neuron into the synapse. Once there, it relays the message to the neuron on the other side of the synapse by binding to dopamine receptors. Once the message has been delivered, the dopamine is dislodged from the receptors and taken up again through a dopamine-specific transporter, the embankments on the neuron of origin.

The circulation of dopamine has the power to send us on a

euphoric trip. It makes us hyperactive, it nurtures our volition and gives us motivation. One of the experiments that established the stimulating power of dopamine involved monkeys and apple juice. When the monkeys were given drops of apple juice after having served in an experiment, their dopaminergic neurons screamed, as proof of their excitement at the treat.[10]

It has now become increasingly clear that the release of dopamine is not associated with the enjoyment of the reward itself. It rather accompanies moments of hopeful anticipation of a reward. Say you are expecting a charming goodnight text message, you have been promised a nice long handwritten letter from a friend, or you may have been invited to a dinner party where you know you'll meet people you like to hang out with or you see the perfect ending to a poem right before your eyes. The prediction of an imminent reward carried by all such promising events is underscored by the production of dopamine. Several experiments have revealed this. When the reward of the apple juice was consistently preceded by the presentation of a light, the monkeys learnt to associate the visual cue with the promise of the juice. The result was that their neurons would fire as soon as the light went on, then not as much when they actually got the juice.

The same was observed in a different kind of appetitive eagerness: sexual anticipation. Levels of dopamine sky-rocketed in male rats lured by the sight of a female kept behind a separating see-through screen. After they copulated with her, their dopamine sank back to baseline levels but surged again at the sight of a second female partner.[11] Such is the power of dopamine-enhanced lust.

But dopamine also helps us focus. It sharpens attention and biases our concentration and actions. To make this clear, I'll tell you a little story about bees. Bees relish pollen. They will travel considerable distance to find a meadow full of flowers. Despite their small nervous systems, bees are quick at learning and processing new information and this helps their foraging abilities.[12] They are able to associate the scent, colour and shape of a flower with the quality of its nectar and this kind of appetitive learning conditions their search for good fields. The experience of finding good nectar involves the bees' reward system. An inviting flower makes their reward neuron produce a bee equivalent of dopamine, called octopamine, and this marks the decision of having chosen that flower as rewarding. It motivates the bee to go back there.[13]

In the bees, this basic system works well enough to satisfy their foraging needs. But in higher animals, it has achieved greater sophistication and the anticipation and detection of pleasure can indeed take us far. The prefrontal cortex (PFC) is the loftiest department of the brain. It's a place where we can entertain abstract ideas, but also a temporary lounge for items in our memory. The reward system is well connected to the PFC and this is important for integrating pleasure with our cognitive abilities. The two parts of the brain process information differently. The reward system is a more crude and immediate type of learner. As we have seen, it is great at detecting and storing rewarding experiences. The PFC learns more slowly and needs exercise. Collaboratively, these two systems catalyse the formation of beautiful thoughts.[14]

Now, in general, a good mood improves solving skills and the creative process. Scientists have probed the effect of positive feelings on the solution of problems and cognitive tasks, including the specific case of word association.[15] In a set of studies, participants were challenged to find connections between words. They were shown a succession of three-word groups and asked in each case to come up within a short period of time with a single word that would fit with all three words in the group. The tasks varied in levels of difficulty. For instance:

MOWER ATOMIC FOREIGN _______

In this case, the correct answer was 'power'. If the participants had been given a small gift before the task, say a candy and other refreshments, or if they had watched a short comedy film, their rate of success at filling the blanks improved.[16] In a related study a small unexpected reward improved the participants' capacity to come up with unusual word associations. Those who had been given gifts were more adventurous in their word associations.[17]

The current biological model in explaining how such mental processes are facilitated credits the importance of connections between the reward centre and frontal structures in the brain. Were I to apply this model to what happened to me as I composed the sonnet, my moment of pleasurable inspiration could be explained more or less as follows. For some inexplicable reason, the new environment I was in, the excitement about being in New York, and even the stars in the sky and the view of the Hudson were the hook for my progress. They were unexpected rewards, sudden incentives, the flowers with the best and most abundant nectar if you want, that carried the potential of something good and demanded that I linger on them. After being stuck with the same few unfinished lines and unfruitful words, my mind finally landed on something promising. Energized and motivated, I didn't let the inspiration dissipate and I concentrated on finishing the sonnet. I sprinted as if I were being chased because I knew I had to bring all the pieces together with renewed poise and a good dose of self-confidence. The rewarding inspiration fitted with the general idea of the poem and therefore grew roots. What had been floating freely in my mind finally found a good landing spot, and somehow I gained access to a productive channel of flexible creation. The ideas were judged worth pursuing and achieved a better organization. Importantly, the regularity and fixed pattern of the sonnet form must have helped in the process. The solutions I came up with found confirmation within a given structure, and were aided by my knowledge of and experience with the sonnet rules.

The ongoing dialogue between the pleasure centre in the

brain and my prefrontal cortex is something I merely now assume was going on in my brain, based on my subsequent reading about the neurobiological infrastructure of the creative process. On the other hand, I can tell you exactly how the sonnet came about because I remember how I built it and my notebook documents the gradual progress, line after line, syllable after syllable, stress after stress. The scribbles and erasures on paper mark the tempo of its production. I remember fondly the ecstatic moments that separated me from the completion, the distance travelled from the suspicion of a solution to the actual realization, and the triumphant high originating in the accomplishment of the sonnet. Knowledge that something specific, and incredibly sophisticated, is happening to my brain while I'm entirely intent on shaping a few lines is extremely fascinating, and definitely reassuring. However, it's an approximation that frames the process, an enterprise parallel to but independent from my own undertaking. What I mostly take from it is that incitements make me a little keener.

## The real thing

Who knows, maybe the sonnet would have been different, or better for that matter, if I had composed it under the effect of drugs.

Creation of all kinds is aided by the use of stimulant and recreational drugs.

Dopamine levels in the brain increase drastically – up to a thousandfold – after the ingestion of cocaine, which vastly amplifies the motivational high I described above. At the molecular level, cocaine increases dopamine levels by preventing its clearance from between the cells in the reward circuit. It inhibits dopamine's re-uptake through the embankments on the pre-synaptic neurons.

Stimulants such as amphetamines work by a similar mechanism and real poets have exploited their power to sharpen focus and boost concentration while letting fatigue dissipate. In post–Second World War New York, the Beat generation of writers that included Jack Kerouac, Allen Ginsberg and William Burroughs experimented widely with amphetamines. In his iconic poem *Howl*, Ginsberg says: 'I saw the best minds of my generation destroyed by madness, starving hysterical naked, dragging themselves through the negro streets at dawn looking for an angry fix . . . '[18]

Speed was a common diversion. Today, the use of amphetamines to enhance concentration and performance is on the rise even among college students and faculty.[19] A survey conducted among over a thousand readers of the scientific journal *Nature* revealed that one in five of the respondents – most of them presumably scientists – made use of some kind of performance enhancers.[20]

I have talked a lot about the anticipation of pleasure, but what is it that pervades moments of ecstasy as they happen? As we have learnt, the anticipation of pleasure and pleasure itself are two different matters. And this difference has been studied at the level of brain tissues and molecules. Broadly speaking, if dopamine is the molecule of motivational pleasure, opioids are the molecules taking care of comforting, blissful sensations.

To make a parallel with a common experience in today's world, dopamine is what supposedly bathes your brain when, having posted something humorous on your Facebook wall, you log on repeatedly in anticipation of unpredictable reactions from your friends. Opioids are probably released when you see the red notifications of each comment or 'like'. The whole thing makes you crave more. Dopamine will make you post something else.

Historically, those who consumed opium mostly retained fond memories of the experience.

Opioids work by binding to dedicated receptors in the brain. In chapter 4, I mentioned how opioids powerfully fight pain. Morphine, for instance, is a very strong painkiller. But opiates also influence pleasure as much as they influence pain. From morphine, through the addition of just two small acetyl groups, derives heroin – smack. Fortunately, we don't need to resort to opium, morphine or heroin to benefit from some of the analgesic, calming and comforting effects of opioids. Our bodies produce their own molecules that resemble opium, bind to the same receptors and work to dampen our sensations of pain. These home-made opioids are called endorphins. In the absence of pain, they are bearers of pleasure and comfort.

As I briefly mentioned earlier, giving opiates to the young of various mammalian species separated from their mothers reduced their protest against the separation. They calm us down. Opiates are also released when somebody simply strokes us. A caress is enough to open the gates to a flood of opioids.

With a few obvious differences, a night of wild oblivious sex, a Beethoven sonata and a succulent meal have much in common, when it comes to their map on the brain. I am going to very briefly illustrate how.

Opioids abound during sex. When we reach an orgasm, the brain looks as if it is on heroin. Much of sex obviously takes place between the legs, but it resonates throughout the body and the pleasure deriving from it travels back and forth between our genital organs and the head. The wires for such communication are nerves which relay sensations of touch and stimulation from our genital organs to the brain via our spinal cord. Long distances are covered by neural highways connecting the brain to parts of the body such as the scrotum, penis, clitoris, vagina, cervix and rectum. The clitoris alone is innervated by thousands of such wires.

In late 2011, at the annual conference of the Society for Neuroscience, an exciting movie was shown to the attending delegates. It was a short clip showing images of the brain during all phases of a woman's orgasm, from its initial approach through to its climax and its fading, a total of five minutes.[21] The clip was presented by the psychologist Barry Komisaruk, working with Beverly Whipple. Komisaruk has monitored the brain's activity in women who successfully stimulate themselves in an fMRI scanner – a remarkable event given the claustrophobic nature of the scanner, but totally possible. At first sight, it looked as though there was in fact no region in the brain that did not light up. Everything seemed to be in ecstatic turmoil. But a closer examination better reveals the neural geography which delineates itself across the timespan of the clip. To go through in detail all the brain areas flooded with oxygen in an orgasm – around thirty on a rough count – would be a boring list, far-flung from the bliss experienced during one. There have been several studies looking at this and some of the results contradict each other, requiring future refinement. However, some brain areas are a constant. For instance, when the orgasm reaches its peak, the reward centre is definitely involved. Of note is the quietness of the orbitofrontal cortex. This being the part of the brain that exercises control over much of our behaviour – Freud's superego – it is kind of reassuring that it should shut off during orgasm, a moment of temporary bliss, oblivious to any kind of mental restraint. Similarly, data coming from inspection of the brain during male ejaculation reveal no involvement of the amygdala. Orgasmic moments bring us to a fearless place.[22]

Placing the orgasm under the scrutiny of science and understanding how it works may help those who have troubles reaching them, such as women with spinal cord injuries. Until not long ago, women with spinal cord injuries were advised to give up on a satisfying sexual life, because everyone thought the severance of the nerves passing through the spinal cord disrupted the pleasure wires. However, Komisaruk and Whipple discovered an alternative orgasmic pathway: the route of the vagus nerve. In Latin, *vagus* means 'vagabond', 'wandering' or 'itinerant'. Indeed, the vagus nerve travels and spans a considerable distance in our bodies. Originating in the brainstem, the 'power-switch' of the brain, more or less at the base of our skull, it departs from the medulla and then weaves itself down the neck along vital paths such as the jugular vein, to then innervate chest, abdomen and our guts. Since the vagus nerve takes the 'guts' route, it bypasses the spinal cord. Indeed, when the injured women stimulated themselves in the brain scanner and reached orgasm, the medulla, which is where the vagus nerve projects in the brain, was active.[23]

'Music is the shorthand of emotion,' said Leo Tolstoy. It is hard to disagree.

In previous chapters, I mentioned how visual art and theatrical performance have the power to elicit strong emotions. Hardly anybody can resist the spellbinding power of music. A pleasant melody, the perfect pitch, a convincing rhythm can be sources of ecstatic pleasure. Why we enjoy music so much remains a mystery. The evolutionary function of music is not evident. In *The Descent of Man* Darwin writes that 'musical notes and rhythm were first acquired by the male or female progenitors of mankind for the sake of charming the opposite sex. Thus musical tones became firmly associated with some of the strongest passions an animal is capable of feeling . . . '[24] So music may have originated in courtship.

Translating the emotional impact of music into words, or into neuronal language, is bound to be a meagre approximation. To feel the rapturous force of music, one just has to listen to it. But imagine you are at the Proms, sitting on the floor with your eyes closed. The conductor reaches his spot while the orchestra prepares. Everybody is waiting for the same thing. Then he lifts his stick and marks the start of the first movement with a slight controlled gesture of his hand. And the first notes, synchronously played, obediently emanate from the strings and travel across the auditorium to kindle you. Whether it is a symphony, a piano sonata or a song, if you appreciate music you may be familiar with the chills, those tingles and shivers you get down the spine or behind the neck beyond your control when you are stirred by music. Nobody really knows exactly why and how such musical frisson happens, but, if anything, it is certainly a proof of emotional arousal in response to music, and a sign of pleasure. The phenomenon was first studied empirically in 1980 and found to be widely common in the population.[25] Chills may be very brief or last for a few seconds. They can actually extend to the limbs and spread throughout the body. Often they are accompanied by piloerection, a fancy word for goose-bumps. Chills seem to occur in response to specific points in the structure of a musical piece. They happen when in the music there are sudden dynamic changes or new and unexpected harmonies.[26] They have also been reported in response to sad music more often than in response to happy music. Opioids partake of these ecstatic harmonious moments and so does the pleasure centre. A group of researchers monitored the blood flow in the brains of people while they listened to music they liked and which gave them the chills. The music chosen by the participants in the study included Rachmaninov's Piano Concerto No. 3 in D minor and Barber's Adagio for Strings. The areas of the brain that were involved as the listeners experienced chills are no different from those stimulated during the enjoyment of food or sex. While the amygdala and the orbitofrontal cortex were quiet, areas of the pleasure centre such as the nucleus accumbens, which are replete with dopamine and with opioid receptors, were highly engaged.[27] Interestingly, if music listeners are given an opioid antagonist – that is, a molecule that prevents opioids from binding to their receptors – they experience fewer chills.[28]

Imagine you have starved for a couple of days. You haven't eaten anything, none of the fancy sandwiches you usually have for lunch, none of the delicious cakes from the bakery downstairs, or the curry at your favourite Indian restaurant with all those spices, not a fruit, not even a piece of bread. Then, when you have sensibly decided to return to food, you treat yourself to a bowl of boiled broccoli which you normally detest. You will eat it, and gladly. Food is weird. It is our most basic fuel, but also a luxury good. It is something we sadly can eat almost distractedly in front of our laptops, but also an indulgence some of us are prepared to spend a lot of money on, if it promises an exclusive pleasure. It is both a bare necessity and a reason for sophisticated satisfaction.

Scientists study this aspect of food by making the distinction between wanting and liking. We want food when we need it. But then we may like strawberries and dislike pineapple. Again, such distinction is mediated by the two components of the reward system, dopamine and the opioids, and this too became clear in experiments with rats. If you shut off their dopamine, rats are still able to distinguish a sugary from a bitter taste, and prefer the former. By contrast, if the opioid system is impaired, the rats will lose appetite and the pleasure-related preference for sweet food.[29] Opioids are needed to appreciate flavours, too. In an experiment rats were presented with two foods to choose from. They were nutritionally identical, but had different flavours, of which the rats preferred one. If then the rodents were given a substance that activated the opioid system, they would go for their preferred food. If instead they were given a substance that shut down the opioid system, they would take either food.[30]

Unfortunately, as I emphasized earlier, pleasure and pain are two sides of a double-edged sword. If abused, pleasure responds with a cold revenge. Whatever at first provided a sense of comfort can later stab you in the back. Opioids increase dopamine levels, thereby fuelling your desire. After repeated exposure to a type of pleasure, your pleasure centre becomes accustomed to it. It is smothered. In addition, after a high comes the low. So, to escape the painful symptoms of withdrawal and to satisfy the increasing desire, you'll just want more and more of the initial reward, whatever that is, but you won't take pleasure in it any more. Drugs interfere with the neurotransmission of dopamine, modifying the structure of neurons in the dopamine system. Addictive pleasures condition your response to cues that remind you of that reward. Even just the sight of the reward sends you on a craving trip. Desire and motivation get out of control.

## Whose side are you on?

A group of neurologists were incredulous when they came across some stroke patients whose brain damage made them exhibit extreme emotional symptoms, but at one end only of the emotional spectrum: either pathological crying or pathological laughing.[31] Those who couldn't stop crying, or cried at inopportune moments, were patients whose stroke had affected the left-hand side of the brain. Alongside their bouts of tears, they also manifested feelings of despair, hopelessness and self-blame. By contrast, in those patients who experienced peals of laughing, the damage in the brain was on the right-hand side. Patients were euphoric and showed elation, a tendency to joke as well as to minimize their own symptoms. Such oddities made neurologists nurture a suspicion: that when it comes to regulating emotion, the brain takes sides. Broadly, the left side is responsible for positive emotions, while the right side takes care of negative emotions.

It's the first time that I have mentioned this peculiar 'handedness' of the brain. As you know, the brain is divided into two identical hemispheres. That means that each of its structures comes as a pair – two amygdalas, two hippocampi, two striata, a pair of cortices and so on – one in each hemisphere. When we speak about the functions of each structure or its involvement in a given cerebral activity, we commonly mean both sides of the brain. But in some cases the involvement is in one hemisphere only. So the hemispheres are identical, but each of them accomplishes a set of different jobs. The most well-known example of a function governed by one side of the brain alone is the faculty to produce speech and to comprehend language, which in most people is the responsibility of the left hemisphere – as discovered in the nineteenth century by the neurologists Paul Broca and Karl Wernicke who have given their names to the particular areas concerned.

The emotional 'handedness' first observed in patients with stroke damage in only one hemisphere inspired the neuroscientist

Richard Davidson, now at the University of Wisconsin in Madison, to explore how brain asymmetry influences the way we emote, even in the absence of brain damage. One of the first studies he conducted was based on an experiment that he recommends you try on your own at home.[32] Stand in front of a mirror and ask yourself a question that needs a little bit of thinking, for example 'What is the antonym for indifferent?' Then, while you formulate the answer, quickly notice the direction of your gaze. Your eyes will move in the opposite direction from the side of the brain that is thinking about the solution. Since questions to do with language keep the left hemisphere busy, in the case of my example your eyes will most probably move to the right. A question about spatial imagery, which is a specialized function of the right hemisphere, will move your eyes to the left.

Davidson employed this charming experiment to probe emotions. When he asked people to recall negative emotions – with prompts like: 'Picture and describe the last situation in which you cried', or 'For you, is anger or hate a stronger emotion?' – their eyes would mostly turn to the left.[33] This confirmed his suspicion that the right hemisphere is in general involved in the processing of negative emotions, but he needed further proof. He needed clear signs from the brain. The best technique available to him was electroencephalography, or EEG. With the help of electrodes applied throughout the scalp, EEG detects with fair precision the fast fluctuations of electrical activity across the entire brain, so you can record which part of the brain is involved during the manifestation of emotions. To elicit positive or negative emotions Davidson used short video clips that either provoked happiness and amusement, or fear, sadness and disgust.

For instance, ten-month-old babies who watched a video of an actress laughing responded with a vigorous smile and had their left hemispheres sparkling with activity. If they watched an actress cry, they would cry in return and in this case electrical activity would traverse the right side of their brain.[34] Similar electrical variation was observed in adults, too. In a couple of other studies, Davidson discovered that left–right asymmetry lay behind the differences in facial expressions corresponding to positive and negative emotions. Happiness corresponded to left-sided brain activity, whereas disgust went together with activity in the right side of the brain.[35] Fascinatingly, asymmetrical brain activity also lies behind the manifestation of a proper Duchenne smile, a smile that involves the contraction of the muscles around the eye. When watching films evoking positive emotions, viewers produced more authentic Duchenne smiles and their manifestation reflected asymmetrical activity in the left part of the brain.[36]

An important implication of Davidson's studies was the possibility that everyone shows different default levels of left or right electrical activity in life, even in the absence of a stimulus such as the video clips he used in the lab, and that these differences influence the way we behave and feel in given circumstances, be they positive or negative. For instance, Davidson found that differences in babies' baseline left- and right-brain activity reflect how they behave in response to separation from their mother. Babies who show higher EEG right-brain activity are more likely to weep and protest strongly if their mothers leave them alone in a room for a short period of time than are same-age babies with higher left-brain activity.[37] Another confirmation of this arrived when he measured the electrical activity in the hemispheres of people who were depressed and whose despondency reduced their propensity to feel positive emotions. People with depression had indeed lower baseline activity in their left hemisphere, compared with people who were not depressed.[38]

But for Davidson, more remarkable than the difference in electrical activity between the two hemispheres in one individual was the difference in electrical activity in the same side of the brain across individuals – say, how two different individuals reacted to the same amusing video clip. In some cases, such disparity was huge. This means that we are all differently equipped to respond to the various circumstances in life. As I explained in chapters 3 and 4, we all react differently to trauma and to loss. The same applies to the way we react to more positive events. We all have different *emotional styles* that are the outcome of a combination of genetic differences, neural circuits and life experience.[39]

You might be wondering: why would the brain use only one side for positive emotions and the other for negative emotions? What is the purpose of such division of labour? Davidson speculates that it might help to minimize confusion in the way we respond to life circumstances. This brings us back to the notion of the fundamental human capacities of approach and avoidance, the strategies at our disposal to juggle pleasure and pain. When we need to shun danger, it would be disadvantageous if our tendency to approach interfered with our methods of avoidance. So, perhaps the brain confided each strategy to only one hemisphere to reduce undesirable mistakes.

## It's now or never

The American intellectual Gore Vidal once told an incredible joke while speaking on radio. It was about a visit paid to the President of France Charles de Gaulle and his wife by the former British Prime Minister Harold Macmillan. On that occasion, Macmillan asked Madame de Gaulle what she eagerly awaited from her future retirement. Apparently, it took the French First Lady no time to say: 'A penis.' At first, the British gentleman didn't know how to react to that startling answer. He tentatively went: '. . . I can see your point of view . . . not much time for that sort of thing nowadays.' Later, Macmillan realized that what his hostess had said, in a heavy French accent, was simply: 'Happiness.'[40]

Whether or not this funny anecdote was based on a real event, Madame de Gaulle's answer voices a widely held attitude towards life. Indeed, when thinking about happiness, it's easy to become long-sighted. Happiness is often regarded as a yearned-for distant trophy at the end of a long journey. We think of it as something we only achieve over time, through endurance, sacrifice and via routes filled with pain and mishaps. We think that happiness is when our lives are sorted, when we have achieved desirable long-term goals and when our circumstances finally coincide with a certain ideal existence that we have constructed for ourselves: say a good job and a devoted partner, or a family, perhaps a piece of property and economic stability, and the prospect of a healthy, carefree existence permeated with all kinds of personal and professional satisfaction. There are certainly no fixed guidelines for an ideal life. Each of us will have our own ambitions. But whatever those may be, the achievement of happiness is a huge driver behind our daily routines, something we know we need to strive for, because it happens later. When somebody asks me the question 'Are you happy?', I often reply with 'Have you got a second question, please?' That doesn't mean that I don't have an idea of what happiness might be. But if I am asked about how I am feeling in a given moment, I prefer to say that I am joyful or that I am experiencing pleasure.

Psychology and neuroscience have not been alone in the search for a definition of happiness and pathways leading to it. Philosophers have been coming up with answers for a much longer time. In their hands, questions about happiness inevitably metamorphosed into ethical questions such as: what is the best way to behave, or how should one live?

Philosophers' ideas on the nature of happiness and how it is to be achieved broadly speaking adopt one of two fundamental approaches. The first of these is *hedonism*. Like most enduring philosophical teachings, it originated in ancient Greece, where it was heralded by Aristippus and later elaborated by the philosopher Epicurus. In essence hedonism is about our most immediate feelings of happiness. It is an invitation to pursue gratification and urges us to maximize pleasure and reduce pain to the minimum. In fact, hedonism resonates with our most basic goal as biological organisms, that of achieving pleasure.

The other fundamental approach to the attaining of happiness is *eudaimonia*, which literally means 'good spirit', but is often translated as 'flourishing', or a 'life well lived'. It has to do with finding and cultivating one's true potential virtues and then living by them. According to a eudaimonic philosophy, there are goods other than pleasure – knowledge, family, courage, kindness, honesty and so on – that are more worth pursuing.

Inevitably, a moral hierarchy has been erected, with eudaimonia gaining the moral high ground. Indeed, hedonism has a bad reputation. This is because hedonistic pleasures are often regarded as ephemeral. They come and go. They are dependent on contingencies and are prone to be unforgivably replaced by pain. As I said at the beginning of the chapter, they are only departures from other less favourable or less pleasurable circumstances. A night of drinks with friends carries the risk of hangover the day after. Eudaimonia, on the other hand, has little to do with fleeting pleasures. It is a better guarantee of stable happiness.

There was one era in history in which hedonism widely came into higher regard: the Enlightenment. Much as the Enlightenment signified the triumph of reason, it was also a fertile ground for the cultivation of pleasure, and happiness. In fact, the enlightened rehabilitation of the pursuit of pleasure had roots in the renewed faith in science. According to nature, mankind shared elementary drives with lower animals, so everyone was born to seek pleasure. Individuals were encouraged to pursue fulfilment, and pleasure was a route to self-improvement.

Earlier, I spoke about bees, their dopamine and their rewarding meadows. At the beginning of the eighteenth century, the Dutch-British poet and physician Bernard Mandeville wrote a long poem that used bees and their capacity to lose themselves in pleasure as a metaphor for human society. First published in 1714, it was entitled the *Fable of the Bees, or Private Vices, Public Benefits* and used bees in much the same way as the ancient Greek storyteller Aesop would use animals to describe human types. In Mandeville's view, the beehive was symbolic of a morally unrestrained society, a collection of individuals each driven by their own competing desires. Somehow, the summation of their deeds, each guided by self-interest, would be beneficial for the entire hive. In his words: 'every Part was full of Vice, Yet the whole Mass a Paradise'.[41] Men and women, however, had the advantage of an intellect that made them select and chase pleasures with measure and sensibility. Pleasure in the Enlightenment was not about excess, but a refined form of self-gratification, the sort of attitude, one could say, that would harmoniously combine both Madame de Gaulle's misheard and actual answer.[42]

In truth, hedonism and eudaimonia are not mutually exclusive. You can shape your character and develop admirable virtues while exercising the ability to enjoy pleasure. Pleasure does not always equate to selfish fleeting satisfaction and can be nurtured by higher goals. It is possible to have hedonistic motivations, achieve momentary happiness, while still keeping an eye on your long-term plans. Fleeting rewards don't come in the way of self-improvement. In short, you can at once be hedonistic and embrace eudaimonia. By avoiding the dangerous drawbacks of pleasure, such as the obsessive chains of addiction, you can exploit the joy derived from your gratifying predilections. For life is short, but it is even shorter if spent unhappily. Basically, you don't have to wait for your retirement to attain happiness. A life spent in anger, fear or guilt is going to be shorter than a life made of joy. Joyful moments add up and build a happier life.

Moments of joy, time spent smiling and laughing and in general in a good mood do have tangible repercussions on our well-being. Their trace can be found in our bodies.

For instance, go and find an old picture of yourself as a child or as a teenager and check if you were smiling. It might tell you how happy you are now. Two researchers in the United States browsed through the 1958 and 1960 yearbooks from a private women's college in the San Francisco Bay area.[43] They were looking for genuine smiles. As I mentioned earlier, if your eyes don't wrinkle, you are probably smiling out of reasons other than joy. Of all the smiles examined in the yearbooks, only half were full Duchenne smiles. The study aimed to find out whether individual emotional tendencies that arise early in life contribute to building people's adult personalities and interpersonal attitudes. To do that, they followed up the lives of the women smiling, for thirty years. It turned out that the women who in the pictures showed clear signs of joy, with a full Duchenne smile, had altogether better lives. They were more caring and sociable. They were also more likely to experience cheerfulness and sympathy. In general, they were less susceptible to recurrent negative emotions. One specific life outcome the researchers looked at was the women's marital status. Those with a proper smile were more likely to be married by the age of 27 and still to be married at the age of 52, reporting satisfying relationships.

A similar study looked at smiling faces in pictures of US league baseball players who had played in the 1952 season. This time, researchers checked whether the presence of a genuine smile could predict a player's longevity. Indeed, those who displayed a Duchenne smile lived on average five years longer than those with a non-Duchenne smile and eight years longer than those who did not smile at all.[44] Eight years is not a negligible margin. It is worth learning to smile genuinely as a child. Renewed contraction of the orbicularis oculi is also a sign of recovery from grief, as observed in bereaved people two years after the experience of loss.[45]

'A day without laughter is a day wasted,' said Charlie Chaplin. There is ongoing debate whether laughing is indeed a universal medicine, a panacea for a good mood. But if a smile can extend your life, there is a good chance that laughter might help, too. If nothing else, laughter can ease situations of pain. Again using a series of videos, scientists have shown that a good laugh raises the pain threshold of viewers. When participants in the study were shown a factual documentary, not much happened, but when they were shown a comic video, viewers who laughed could better sustain the pain of a tight cuff around their arms or contact with a frozen wine cooler sleeve.[46] The effect was stronger when the viewers laughed in a group rather than when they watched and laughed at the video alone. Behind the raised pain threshold is the release of endorphins.

In general, a positive disposition does improve physical health. Feeling calm, cheerful and strong as opposed to sad, tense or angry can even increase your resistance to developing a cold![47]

## Down to a single nerve

Whether or not you are, like me, uneasy with being asked if you are happy, psychologists have learnt how to quantify happiness. Typical surveys of happiness explore whether, all things considered, people are satisfied with their lives, and to what degree, or if instead they would like to change anything.

In his book on the science of happiness, the economist Richard Layard talks about seven main factors that contribute to happiness: health, employment, income, freedom, personal values, family, and social relationships and friends.[48] Of these, against common thinking, money and our financial situation are, in truth, the least influential. Happiness does not necessarily increase as a consequence of higher income. Rich people are not happier than the poor. Surveys have shown that, once what we earn has covered our basic needs, the surplus money doesn't buy us happiness.[49] If anything, better incomes make people desire even more.

What does seem to make a difference in levels of happiness is how we choose to *spend* money. Especially whether we pour money into selfish expenditure or whether we use it more altruistically. In the US, a group of about six hundred people were asked to report their happiness and how much they earned. Then, they were asked to list how much of their monthly earnings was on average spent on bills, on gifts for themselves and on donations or gifts for others.[50] The happier bunch were those who had spent more on others. Similarly, when a group of employees rated their happiness before and after receiving a bonus and reported how they had used it, they were clearly happier if they had spent it on things like presents for others, donating to charity or meals with friends rather than goods for themselves. How they spent their windfall meant more than its amount and being generous towards others was a significant factor in the realization of their well-being.

Taking a step back from our own concerns, and reaching out to others and embracing theirs is usually a source of happiness. A self-effacing attitude may earn us amplified rewards.[51]

From time to time, especially after a chaotic day spent running around people we don't know or commuting in packed tube carriages, we may appreciate going solo, relishing the luxury of withdrawing from the world and enjoying the peace of solitude. But research on happiness is clear: we are better off when we are not alone. Of all the factors influencing our emotional well-being, by far the most significant is the establishment of social and emotional bonds. To be circled by people is good enough. It is even better if we surround ourselves with people with whom we have meaningful relationships. So, thousands of Facebook friends don't count very much unless they are all good and dear friends.

Satisfactory social relationships improve the quality of life and considerably extend longevity, too. A systematic review of mortality studies on about 300,000 individuals across the world showed that people with satisfactory social relationships improve their chances of survival by 50 per cent compared with those with poor or inadequate relationships.[52] The effect of having good friends is almost equivalent to the effect of quitting smoking and is greater than that of either physical exercise or abstinence from alcohol.

Friends have the capacity to uplift us and our relationship with them seems to affect us deep under our skin. If positive emotions have beneficial effects on our body and our health, it should be possible to discover physical indices of such improvements.

In search of such clues, psychologist Barbara Fredrickson has found one that is measurable at the level of a single nerve, the vagus nerve. Brains have long tails. Earlier, I mentioned how the vagus nerve is involved in achieving orgasms. It seems that it is also of help when we engage in social interactions. In general, the vagus nerve acts as a communication device that senses how our main organs are doing and sends this information back to the brain. One index of whether or not the vagus nerve is functioning properly is called the cardiac vagal tone. It reflects the variability of our heart rate during respiratory performance. Even if we can't perceive it, our pulse is slightly more rapid when we inhale and slightly slower when we exhale. The vagal tone corresponds to the amount of difference between these fluctuations.[53]

Fredrickson has established how cardiac vagal tone is a signature both for our physical health and also for our propensity to feel positive emotions and that the two are, in fact, connected. A high vagal tone gives you the capacity to take advantage of positive circumstances. As Fredrickson puts it, it gives you the chance to capitalize and expand on your positive emotions in order to build, through the additive value of positive moments, rich personal resources that amplify your well-being. This is facilitated by the establishment and appreciation of social connections.

In one experiment, Frederickson and her collaborators monitored for nine weeks in a row the vagal tone and the emotional well-being of a group of individuals in relationship to their daily social interactions with friends and dear ones.[54]

Those who had a high vagal tone from the start showed rapid increases in social connectedness and reported the experience of positive emotions such as joy, love, gratitude or hope. At the same time, those improvements in social connectedness and positive emotions also predicted increases in their final vagal tone, which was higher at the end of the study. Basically, what the study found is that as we work on our close relationships and instigate social contacts with others, we regulate our cardiac vagal tone which, in turn, backs up and stabilizes our positive emotions. A perfect reciprocal deal between our physical and mental well-being. As a follow-up to the above study, Fredrickson extended her research, asking whether it is possible for people to deliberately work towards an improvement of their vagal tone. Her strategy to generate positive emotions was a meditation technique that induces feelings of love, goodwill and compassion for oneself and others.[55] In combination with the meditation technique, higher vagal tone facilitated the improvement in the perceptions of social relationships and in the manifestation of positive emotions, which in turn increased again the final vagal tone.

What fascinates me about these studies is how slight but meaningful changes in our nerve physiology contribute to influencing our social behaviour. Interestingly, the branches of the vagus nerve are such that they are connected to the muscles governing our facial expression, eye gaze, as well as muscles in our middle ear that sharpen our ability to tune in to the frequency of human voices. Therefore, a positive vagal activity equips us with all the necessary qualities to engage in social behaviour.

So, it does make sense to invest in meaningful friendships and social interactions to contribute to your own and other people's well-being. Darwin once said: 'A man's friendships are one of the best measures of his worth.'

What all this means is that while we steer towards our ideals, and our ideal life, we can enjoy the path. While we pursue a distant happiness, we can exercise skills, pleasures and virtues that can actually help us reach our goals and perhaps shorten the route.

## Coda

Rising early has tangible benefits. I had a chance to savour the small triumph of the finished sonnet and abandon myself into a short state of bliss, without thinking about much, just taking in the sounds and light of the early morning. There are precious pleasures to be enjoyed on a morning walk. To be greeted by the joggers who cross your path, to smile at strangers and pick the person with whom to trade the first words of the day, to meet a parade of strolling dogs, grab the freshest bagel, collect the newspaper for the neighbours. When confined within the close boundaries of repetitive habits, we become in a sense blind to our surroundings. Our mental gaze is projected on to a distant purpose, and we overlook local opportunities of delight. But joy, or even just a small pleasure, gives us better eyes. For joy is also skilled in something else. It nails down fear. It pushes it down into temporary oblivion, vigorously, so that we can make room, look at everything with renewed optimism. Joy has the ability to cultivate itself, if we let it. If I find a reason to be joyful, however small the pleasure is, new joy will be making its way to me by some shortcut – I don't know whether that shortcut is the vagus nerve or another path.

There is another trick to cultivate joy that I am fond of. In 1962, the American author James Baldwin published a beautiful essay entitled 'From a region in my mind' in the *New Yorker*, in which he wrote about the conditions of blacks in America. In a paragraph dedicated to the power of jazz, he wrote how only black people truly know the depths whence it comes. In the middle of it lies this treasurable sentence: 'To be sensual, I think, is to respect and rejoice in the force of life, of life itself, and to be present in all that one does, from the effort of loving to the breaking of bread.'[56] Here the word sensual, as Baldwin acknowledges himself, has nothing to do with the meaning most people associate with it. I interpret the skill of being sensual as the ability to own your actions and fill them with meaning and value, without letting them just occur to you, as if you didn't believe in them. Baldwin's exhortation is a tough call, but also one of great promise. It has haunted me ever since I first read it, but it has also been a source of hope and strength, a reminder to which I can resort whenever I need. What else is there to do but to participate fully in each of our ventures?[57] If I write a line or two, if I scramble eggs, paint a wall, hang a picture, play the piano or do the dishes, I want to do full justice to those actions. Equally, if I dedicate time to my friends, listen to their stories, buy them a present or help them in one way or another, I want to fully enjoy and believe in the generousness of those gestures.

One could even say that, in a nutshell, Baldwin's sentence unifies both the hedonistic and eudaimonic precepts. It helps you find what you most like doing and most believe in, and exhorts you to savour it, master it, cementing the pleasure derived from it to build for your preferred future. It takes courage and determination to find what that is, it might be scary at first. But remember, fear and bravery are two sides of the same coin. And if you practise joy and let it happen to you, courage will emerge. Whistle to keep up the courage, as William James would recommend.

My minor creative achievement in the New York dawn didn't itself need much acclaim. But I knew that a new round of frustration with writing would be just around the corner, so, before the next bout of pain, the temporary joy I was in was not to be ignored. I felt like sharing it.

I learnt the importance of good friendships and conviviality as a child, when family friends in small or large groups would regularly ring our doorbell for company, even late in the evening, and my mother would improvise quick meals to feed the multitudes. 'It's us!' they would shout from the other side of the door. Over time, she invented a pasta dish that became the regular food for those occasions. It was called simply 'Spaghetti my way'. They would embark on conversations on all kinds of subjects, from the latest political issue to the newest film or book, or small local events. Daily successes and failures would be shared. Plans for joint holidays would be plotted. It didn't really matter how the evening would unfold. What mattered was to be spending time together. Music was a constant presence at those gatherings. The piano would be opened wide for those who wanted to play and the adventurous would sing. Everyone was cheerful. Such impromptu visits to the house were enormously entertaining for me and a reward for all of us.

At the end of my morning walk, before returning to the flat, I stopped at the grocery for some food shopping and sent a text message to a bunch of friends: 'Dinner tonight. Come early and we'll cook together!' I would be making the dish I remembered from those childhood evenings. My own way.

# 7

# Love: Syndromes and Sonnets

Love is a better teacher than duty

ALBERT EINSTEIN

I . . . profess to understand nothing but matters of love

PLATO

I recall it all started on a Sunday afternoon at the beginning of April, my second year in graduate school in Heidelberg, south Germany.

Together with a few friends from the lab I had planned to go on a long bike-ride along the river. But spring was late and the weather was uncertain that weekend. Clouds came and went and an annoying drizzle convinced us to change our plans. A movie followed by dinner seemed like a worthwhile alternative, so we all arranged to meet in front of the cinema in time for the six o' clock screening. As often happened, I arrived early, so I waited outside by a stall where they sold ice-cream and popcorn, watching people walk by.

The rain had briefly stopped and, at a certain point, as I turned to see if my friends were coming up the road, my eyes landed on something I would not easily forget.

On the other side of the road, with one leg bent against the wall and an arm embracing a big cello case, stood a tall, dashing young man whose presence shone bright and demanded notice. He looked like he was waiting for someone as well. Our eyes locked and my perception of the surroundings became murky. I felt quickly absorbed into another reality. A building could have collapsed next to me without my noticing it. I read delight in his face as well as a tinge of surprise and curiosity. Then he smiled, complacently, and I smiled in return, as if to acknowledge something we were both eager to discover and know better. Simply paralysed, I pondered what to do. I didn't want to look too eager, but I couldn't keep my eyes off him. I wanted to get closer, find out his name, inspect his face more closely. Was he a tourist, a touring musician, or was he too a student? If so, how was it possible that I had not seen him in town before?

As I asked myself these questions, the boy started to walk towards me. Incredulous, I closed my eyes and lost my breath for a moment. I rushed to come up with something kind and smart to say, but as he was getting closer, it all stopped. 'Here we are! Sorry for being late!' By pure coincidence both his and my friends showed up at the same time and shyness got the better of us. Result: neither of us said anything. The unknown beauty and I looked at each other again briefly and, as I was ushered into the cinema, he and his friends walked off towards the main square. The undying brightness of his face stayed as a blinding after-image in the darkness of the theatre, still against the motion of the movie, lodging itself firmly in the depths of my wishing well.

To this day, I still can't recall what movie I watched. It didn't matter.

The whole of my attention was unreservedly concentrated on the vision of this new creature I had never seen before in town. When I came out of the cinema, there was obviously no trace of him, and I started to obsessively wonder if I would ever see him again.

For a few days, it felt as if I had a fever. My mood swung sharply during the day; I found myself daydreaming, but also restless. I couldn't sleep. I would often think of him, the more so because I was afraid of forgetting his face and what he looked like.

Love is, above all, insanity. When in the initial fiery stages of love, we enter a space in which fears, desires and outlook on life are shifted. Priorities change. The ecstasy taking possession of us is so strong that as well as falling in love with a particular individual, we tend to feel at harmony with the entire world. We become optimistic and overlook things that used to annoy us.

If the target of our desire is the recipient of unfaltering consideration, he or she is also a source of recognition for us. As we highlight and underline the other's attractive properties, when we approach them we are also in search of confirmations of our own value. We enjoy being noticed and appreciated. What we need is gratification, a regard for our own self-worth. Love definitely sits on the positive end of the rainbow of emotions. It is – mostly – a source of joy. Of all emotions, love is perhaps the most complex, ambiguous and unpredictable, but also one of the most rewarding, both when giving and when receiving it. Alone, it encapsulates feelings of joy, anxiety, jealousy, sadness, and even anger, guilt or regret.[1] Almost everyone is or has been interested in it during a phase of their lives, or at its mercy. In 2012 'What is Love?' was the most searched question on Google.[2]

In the previous chapter I concluded that meaningful friendships rank very high as a contributor to happiness. Yet for many, love, by which is meant the reciprocal affection and passion between two individuals, beats friendships. We could all be fine with friends. Yet we seek the exclusive affection of one individual. Though hard to define, and sometimes even more difficult to achieve, true love remains one of the ultimate life goals to which a lot of human beings aspire.

## What's neuroscience got to do with it?

Up until the second half of the twentieth century, molecular explanations of love were not the most prevalent. In our cultural imagery, the fabric of love is not made of molecules and units of DNA, but of passionate, fugitive moments of ardour and union. Love and its secrets also belong to intimate conversations. It seeps through confidential chatter among friends and lovers of all levels of expertise who share their successes or failures in dealing with it, always in search of rules and precedents that can teach how to go about it.

So, the question is: can love be studied in the laboratory and trapped in a test tube? Indeed, from a neuroscience perspective, love is still only sparsely understood. Neuroscientists have the curiosity and ambition to dissect the wonder of love into its neural components. An increasing number of studies involving genetics, neurochemistry and brain imaging have sought to explain all phases and kinds of love, from the passionate establishment of romantic bonds to sexual pleasure, maternal love, relationship attachment and the desolate experience of rejection. Doubtless, this mighty emotion reflects considerable and tangible changes in our bodies.

The fact, for instance, that we focus our attention on one human being alone and that we imaginatively build sexual fantasies, scenarios of intimacy and prospects of union with them reflects enormous changes in our cognitive and emotional life, which, of course, involves tremendous rearrangement in neuronal wiring.

However, especially during the initial phases of my infatuation, my knowledge of neuroscience and experience in the laboratory had little or nothing to offer to make sense of what was going on or what I was feeling – except that I knew my brain was definitely orchestrating the production of more hormones than normal.

You may be wondering: did I find him, or ever see that beauty again?

Of course I did, and relatively fast. Love is an incendiary passion, but also a powerful motivator. I embarked on a resolute mission to find him. I returned to the city centre and the area around the cinema a few times hoping to bump into him again. I asked friends for any clues, roamed all the libraries in town, carefully screened all the bars I went to. And, of course, sieved all the classical music concerts to find the cello again, in case he played in the town or university orchestra. All that fuss for a once-seen man!

Eventually my persistence and incessant search proved fruitful. Unexpectedly, of all places, the stranger appeared again in one of Heidelberg's open-air swimming pools. Who would have foretold? I remember I had been swimming for an hour already and that I was ready to leave, but when I saw him emerge from the changing rooms, I obviously decided to stay longer, with a strong determination to talk to him.

It took one more mile of front crawl, but in the end, we settled on a date.

The madness did not recede. If anything, it increased, unearthing a thin edge of anxiety, too. The day of our meeting, I was electrified. Like I explained in the previous chapter, the expectation of pleasure and reward is already a generous source of well-being. In Germany this is common knowledge: *Vorfreude ist die schönste Freude*, they say: Anticipation is the best joy. It brings excitement. Like a bee finding the best garden in which to forage, I felt I had spotted the best flower.

In the midst of all of this and eager for good advice and tips on how best to behave I left the lab early in the afternoon and plunged into Plato's books on love, convinced that I would find inspiration in those pages.[3] To my good fortune, the ancient Greeks could actually tell me a great deal about some of the dynamics of love, even in the twentieth century. In *Phaedrus*, Plato offers a clear idea of the madness of love. He grants it divine origins and a favourable, important role in our lives. As a divine gift, love can only generate good and makes us search for goodness. It is ranked alongside the experience of being possessed by the Muses of poetry, a 'Bacchic frenzy' – that is a madness similar to being drunk or on a high – without which no poet can, on the basis of linguistic erudition or craft alone, compose any good poetry.

Even nobler than the madness inspired by the Muses of art and poetry, the kind of divine possession felt by a lover is a madness manifested when we see or are reminded of true beauty.

Plato uses an apt image to visualize the condition of love. Love is so exhilarating that it makes us desire to spread wings and rise up. We want to fly high. Unable to do so, we are set into some kind of unremitting motion; we flutter and quiver and 'gaze aloft', wanting to elevate ourselves, and this makes us look as if we had gone mad. The Athenian philosopher also discusses what makes a lover 'successful'. What measured combination of skilful conversation, wit and charm should one employ to best seduce and conquer the beloved? And does it make sense to love someone who doesn't requite our passion? Love is a relentless impulse that generates an inner struggle. To exemplify this tension, Plato used an allegory that has by now become widely celebrated. He said that the mind (in his words the soul, or *nous* in Greek) is comparable to a charioteer driving a pair of winged horses. One of the horses is noble, of good nature, docile and obedient. The other, of opposite bloodline, is irrational, undisciplined and harder to tame.

The allegory is appropriate for matters of love. Charged with poetry and philosophical authority, the image used by Plato reflects a central dilemma in the protocol of love that has persisted throughout time and still haunts lovers nowadays: shall we follow our instinct to seek pleasure – including the pursuit of bodily consumption – or shall we let reason and judgement control our actions? Applied to the early phases of love and courtship this might read: is it helpful to let madness get the better of us, or is it wiser to save our best sentiments for when we are certain that we have conquered him or her? In modern terms: shall we play hard to get, or shall we take the initiative?

A common misunderstanding of Platonic love is that it is entirely void of erotic expression. Love, according to Plato, brims with desire, initially for physical beauty. But this desire evolves and matures. Over time, it will free itself of the tyranny of the senses and will contemplate other, more elevated forms of beauty such as personal and moral beauty, even if trapped in an ageing body. Ultimately, love will ascend to its highest stage, comparable to a savant's passionate quest for and acquisition of knowledge.

Love becomes shared and mutual exploration and can produce beautiful sentiments and ideas.

On the evening of our first date, the stranger and I set our imaginations into motion. We envisioned entire scenes of our immediate and lasting future. We would be having dinners together, go to exhibitions and travel to an exotic destination to mark the start of an enduring relationship. We would work and create together. We also dreamed of evenings on the couch, of expeditions to the farmers' market, hikes on the local hills, a road-trip among the vineyards, endless conversations and mutual entertainment. We had no doubt that together we would discover the highest form of love and that we were sanctioning the start of what we thought would bloom into a perfect relationship.

## Prime sight

What a lasting effect that stranger had on me, or what a powerful 'external stimulus' worthy of approach he was, I should say. I had only seen him for less than five minutes and I decided to pursue him. As George Bernard Shaw put it: 'Love is a gross exaggeration of the difference between one person and everybody else.' A few glances have the power to induce an overwhelming mental and physical response. Can we fall for someone we only briefly saw and about whom we know almost nothing?

Sight is traditionally primal to love and poets have endlessly emphasized its essential role in directing the trajectory of Cupid's arrow. In Ovid's *Metamorphoses*, Apollo, the god of light, after seeing the nymph Daphne is aflame with love for her and pursues her, even though she has no interest in him. After sneaking into a gathering of the Capulets, when Romeo sees Juliet for the first time he instantly falls in love with her and says: 'Did my heart love 'til now? Forswear it sight / For I ne'er saw true beauty 'til this night.'

Apollo, Romeo, and I myself at the entrance of the cinema theatre, seem to be at the mercy of an erratic, whimsical force that ignites our passions beyond our control. It is no surprise that

Cupid, the god of love and the son of Venus and Jupiter, is represented as a child, who arbitrarily shoots his arrows to match two people, almost at random. With or without Cupid's help, how do we get struck by one specific person and not another? Consider the situation of a party. If we are open to finding a lover, the first thing we do when we enter the crowded room is to scan it quickly to identify and focus on the person we consider a possible match for us.

Long before neurology came into the picture, the science of optics inspired poetic representations of love. For poets at the vibrant court of the great Holy Roman Emperor Frederick II in thirteenth-century Sicily, the ignition of love resided in an optical incident. In his court, where scientists and artists of all kinds gathered, Frederick had a talented notary and poet, Jacopo da Lentini (1210–60), who is generally credited with the invention of the sonnet, his preferred form when writing about love. As I mentioned in the previous chapter, I too love sonnets and it is a nice coincidence that I grew up not far from Lentini, Jacopo's home town and also the birthplace of the sonnet. One of Jacopo's most famous sonnets goes like this:

> Love is a desire that comes from the heart
> Through an abundance of great pleasure
> The eyes first generate love, and the heart gives it nourishment
> . . . For the eyes represent to the heart the image
> Of each thing they see, both good and bad . . . [4]

Today, we know that the principal organ of love is not exactly the heart. The arrow of love, whatever this is, pierces the eye and from there it penetrates deep into the brain to the thalamus, where the visual message is processed and then passed on to the fusiform face area. When we meet another human being, the face is usually what we give most of our attention. A face gives away crucial clues to a person's emotional state. The brain regions specializing in face recognition are all connected to the amygdala and to the prefrontal cortex, the two modulators of our emotional experience.

Indeed, many of the studies which have attempted to investigate romantic love have consisted in showing lovers in a brain scanner pictures of their beloved. You definitely cannot re-create the overall experience of a romantic encounter inside the scanner, but you can try to observe how a visual input arouses and sustains an emotional reaction in a person who is madly in love. In 2000, Andreas Bartels and Semir Zeki from University College London asked a group of young volunteers who declared themselves to be intensely in love to participate in a study that investigated the neural systems of romantic love.[5] During the scanning procedure, all participants viewed colour pictures of their loved partners who had requited their feelings for an average duration of a little over two years. In another similar study, Arthur Aron, Helen Fisher and colleagues from Rutgers University, New York, recruited an equal number of participants who also declared themselves to be madly in love, but they had been gripped by that sentiment for a maximum of seventeen months and, therefore, were in an earlier stage of a romantic relationship.[6]

Alongside the brain activity measurements, all participants also ranked their romantic feelings by completing questionnaires that quantified their passion. They were asked to rate statements such as: 'X always seems to be on my mind', 'I possess a powerful attraction for X', 'I yearn to know all about X', or 'I feel happy when I am doing something to make X happy'.[7] These may sound like common questions, but they do work for psychologists to assess the level of passion in lovers. The two studies dovetailed each other, revealing similar results. The areas of the brain that showed the highest activation were primarily two regions below the cortex. One is the ventral tegmental area, which covers the brainstem. The other is the caudate nucleus, a C-shaped structure at the centre of the brain, sitting astride the thalamus, and so named because it has a wider frontal part and a thinner tail – in Latin, *cauda* (Fig. 16). (The nucleus accumbens, another subcortical region, was also involved.) As I described in the previous

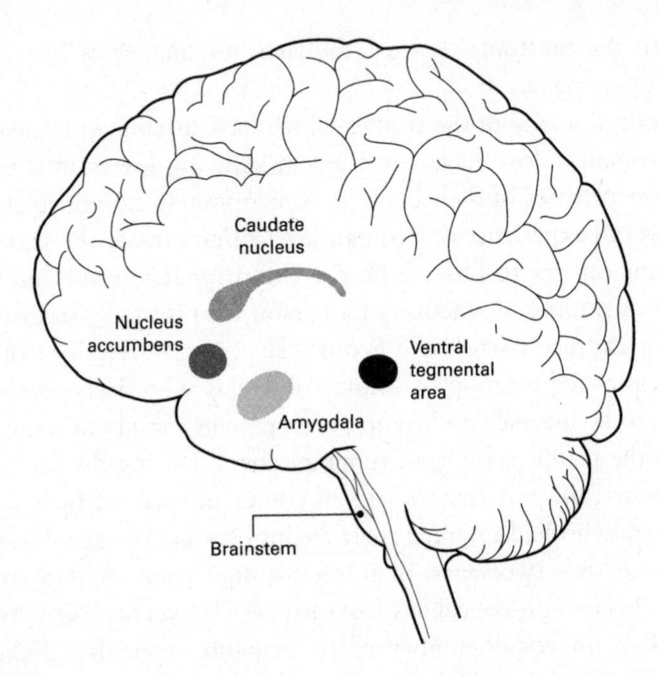


Fig. 16 Brain areas at work when gazing at the picture of a beloved in an fMRI scanner

chapter, all these regions primarily mediate reward and motivation and are bedewed with dopamine to awaken desire. Both the VTA and the caudate nucleus are also well connected to the visual system.

Anyone who has ever had a crush will recognize the dopamine-related behaviour that pertains to it. The hyperactivity, the incredible motivation, the lack of fatigue – I definitely spent many sleepless nights writing poetry inspired by my infatuation.

When dopamine circulates in the brain, our minds are assisted in focusing our attention. Thus, when we are in love, dopamine makes us focus primarily on our beloved. Our thoughts are concentrated on one person. We can't think of anything else. He or she ranks at the top of our priorities, and everyone else around becomes irrelevant or at any rate not as important. Such fixed and exclusive attention allows us also to concentrate on, and remember, details about our targets of desire. We remember what they wore, the exact words they said, we are able to describe the restaurant room where we had dinner with them, and their facial expression when we parted from them.

The fMRI images also revealed deactivation – a decrease or loss in activity – in the amygdala. The amygdala is central to our emotional life and the main repository of our fearful reactions. It is not surprising that during intense, but not so early, phases of romantic love, the sight of a beloved would result in less activity in this area, because the feeling of elation, trust and protection deriving from such a vision is likely to dissipate fear.[8]

## The low after the high

The unquestioned and unquestioning admiration for the object of our desires may not last indefinitely. Nor may the madness that characterized the nascent phases of passion. Over time, when we start to think clearly again, the beloved appears in a different light and may shed a misleading disguise. We shyly wonder: what was all that about?

After months of fervour and reciprocal admiration, the relationship with my boyfriend underwent a few considerable changes. It's not that we grew tired of each other, but we came to discover features in us, and in the way we loved each other, that no longer inspired us. I won't bother you with the details, but slowly I discovered a few unattractive edges in his attitude towards life. Sad as it can be, on a few occasions, it seemed that part of who he was had nothing to do with the person I thought I met on that April Sunday, and with whom I shared a few truly rewarding and affectionate moments. All in all, we realized we were not ready to embark on a search for the best way to live together. What changed, or what did we overlook when we first met?

We have heard this many times: love is basically blind. Not only is love blind, it has a lot of imagination. Inspired by an intense, but unrequited love experience with Mathilde Dembowski, whom he met in Milan in his thirties, the French author Stendhal (1783–1842) wrote a book on love, a large part of which is dedicated to the role of imagination. He compared falling for a person to a natural phenomenon which he named crystallization.[9] If you leave a stick for too long in a salt mine, when you take it out again, it will be fully covered with crystals and will assume a completely different aspect. It will no longer look like a stick. When we are attracted to someone a similar process occurs. We go ahead and imagine. We paint moments of happiness and harmony which nothing can guarantee will in reality take place or be sustained. Not only that. We decorate our object of desire with ornaments. Often, these embellishments are qualities we lack ourselves and that we would like to possess. This is not surprising. We are rarely attracted to something we already have. When I first met him, and during the early phases of our dating, the man whose face had struck me so powerfully looked simply flawless. When in love, our cognition is so disoriented we get euphoric at the merest glimpse of the qualities we long for. If we are eager to possess a great sense of humour, even an average joke made by our beloved sounds like a first-class stand-up set. An occasional beautiful scarf is a sign of timeless elegance and taste in clothes. An assertive remark about their beliefs and convictions is regarded as admirable solid self-confidence.

Interestingly, this recurrent aspect of love showed up in one of the brain-imaging studies that involved looking at pictures of the beloved. Significant neural deactivations were observed in some parts of the brain involved in the processing of negative emotions, the formulation of judgements towards others, as well as the perception of self in relation to others[10] – comparable to the neural changes observed during the suspension of disbelief in theatre watching.

In chapter 2, I explained at length how cautious we need to be when interpreting fMRI results and how difficult it is to allocate attributes of an emotion to distinct brain regions. In this particular case, trying to capture a sentiment as complex as romantic love in a brain scanner does sound like an incredibly ambitious, if not naive enterprise, diminishing the grandeur of the sentiment. However, on the face of it the silencing of these brain areas in states of romantic love would make sense for several reasons. Especially in the nascent phases of love, it is hard to make unbiased remarks about our objects of desire. We don't seem to notice undesirable attributes in them. If we do, we don't give them serious weight, or think that they could worsen; we can foresee growth only in their good qualities. If we express any judgement, that is mostly of a kind and complimentary nature. Basically, impartial judgement vanishes. The French philosopher Roland Barthes (1915–80) compares the lover to an artist whose world is 'reversed', 'since in it each image is its own end'.[11] As if, sadly, in love there were nothing beyond the image. The beloved becomes a ghost, a mere artefact of the imagination.

Secondly, one of the most recurrent feelings in love is that we and those to whom we are attracted reach a strong and self-effacing unity of body and mind. This unity narrows physical and mental distance and, as our trust in the other grows, would also put aside any stark doubts about our sharing their beliefs and ideas. During love, we lower our barriers and defensive strategies. Of note in this regard is that some of the deactivations observed in measurements of romantic love show a close anatomical overlap with deactivations in a region of the frontal cortex observed during sexual arousal and orgasm. Sexual union is, after all, as near as humans can get towards the union of mind and body to which we aspire in romantic love.[12]

## Sight betrayed by emotion

So, early romantic passion can be a long, deceiving afterimage.

A bizarre neurological syndrome known as the Capgras delusion is a particularly intriguing example of how our sense of sight is betrayed by emotion. Patients with the Capgras delusion are otherwise lucid, but they regard a close acquaintance – usually someone with whom they are intimate – as an impostor. The syndrome was first reported in 1923 by Joseph Capgras, a French physician who wrote about the remarkable case of a 53-year-old woman, a certain Madame M.[13] Madame M had reported her husband's sudden disappearance to the police. In truth, her husband was waiting for her at home. Madame M, however, had become convinced that the man she lived with was not her real husband but only a double who looked exactly like him and had stolen his identity. Over time, Madame M continued to experience similar delusions and fabricated a whole new reality for herself. Across a timespan of about five years, she reported having met thousands of unfamiliar messieurs – as she called them – who claimed to be her husband. Each was the double of the previous one and Madame M found something unfamiliar in each of them. Dr Capgras described Madame M's imaginative condition as 'chronic systematic delirium' which he suspected had something to do with a misinterpretation of visual information.[14]

Since Madame M, numerous similar cases have been reported. In a few, symptoms manifested as a consequence of brain injury. Studies have indicated that patients affected by the syndrome can recognize the faces of those they love, but are unable to experience any of the emotions that would normally arise from their familiarity. Simply recognizing somebody and experiencing an emotional connection with them are two different tasks within the brain. In broad outline, the former is mediated by an area called, appropriately, the fusiform face area. The latter is processed in the amygdala where our emotional memories are created and stored. Neurologists suspect that the Capgras symptoms might result from a specific disconnection, or miscommunication, between these two functionally different parts of the brain. Interestingly, the specificity of this missed link is confirmed by the fact that, in the physical absence of the presumed double, the patient can emotionally recognize their real partner – if they hear their voice on the phone, for instance.

The Capgras syndrome has fascinated many.[15] When I first heard about it I became interested in using it as a prism with which to examine love.[16] For how many of us have never had the experience that we stop recognizing those we are attracted to, if not in a literal sense, then emotionally? After all, the image we cherish of those whom we think we know so intimately can sometimes turn out to be distant from reality. As we discover new and unexpected faults that we never noticed before, those whom we love can gradually begin to feel like strangers to us. In effect, they become impostors.

Why does love change? Is it because we or our beloved constantly change, or because our sensory perception is betrayed by our emotions? Or is the reason simply that our eyes constantly need novelty to sustain our desire?

In the Capgras syndrome, nothing is apparently wrong with the visual perception of the beloved *per se*. The problem lies in the *interpretative* aspect of the visual recognition – in other words, in the judgement we make of the visual information through our emotions.

In his Sonnet 148, William Shakespeare evokes the antagonism between sight and emotional judgement:

> O me, what eyes hath love put in my head,
> Which have no correspondence with true sight!
> Or, if they have, where is my judgement fled,
> That censures falsely what they see aright?

Exploring the Capgras syndrome offers insights into one of the fundamental aspects of love: the disparity between the person we think we fall in love with, and who they really are. Such disparity may also reveal itself between lovers who have been together for a long time, but it originates critically in the initial stages, when the euphoria of love may make us fabricate an entirely distorted projection of the other, based on the person we wish and idealize our partners to be.

One problem is that we often want to go back to the experience of the passion of the early days. We long for the feverish, incendiary love condition of the start. We wish we could for ever be like Romeo and Juliet.

Romeo and Juliet are the epitome of enduring romantic love.

The two young Italian sweethearts did not have the chance to regret the end of their passion for each other, because they died before their sentiment could wilt or dissipate.

Every year, Valentine's Day is a ritual by which lovers commemorate their mutual romantic feelings. For those who have been together for a while, it is an excuse to reignite the euphoric passion of the early days of their relationship. Long-term lovers are well aware that they can reinvigorate attraction by introducing novelty – whether through sex, or by changing hair styles, wearing new clothes, buying flowers or in general surprising the other. By doing so, they are teasing the dopaminergic neurons, satisfying their need for novelty. They are turning the old into new. In the case of the Capgras delusion, any attempt to employ the stimulation of dopaminergic neurons to strengthen or revive attraction would mean to actually let the new seem old.

## Winds of commitment

So far, I have talked about love as a kind of infatuation which may reveal itself as an illusion. But boundless longing – and even crystallization – does not always end in nothing. It may mature and develop into something else.[17]

What keeps a relationship going over time? What cements a bond after the early passion?

If Romeo and Juliet's boundless love had continued, it most likely would have taken the regular course of any other relationship. I am not saying they would have ended up hating each other or separating, but their bond would have most probably matured into a form of attachment very different from the ardent, magnetic attraction of their first encounter. Two hormones, oxytocin and vasopressin, have been hypothesized to play the principal role in the more mature phases of love, in the quiet whereabouts of long-term attachment.

Oxytocin and vasopressin are small hormones called neuropeptides that are produced in the hypothalamus but are projected and function in other parts of the brain by binding to receptors.

The molecular confirmation of the role of these two hormones in influencing attachment was discovered, believe it or not, in voles, also known as field mice.

Two species of *Microtus* vole display remarkable differences in their respective behaviour. Voles of the species that lives primarily on prairies (*Microtus ochrogaster*) are highly sociable and monogamous. Husband and wife spend most of their time together, are jealous of their partners, and they also cooperate in taking care of their offspring. Mountain-dwelling voles (*Microtus montanus*) on the contrary are extremely antisocial and promiscuous. They engage in 'extramarital' sexual activities and often neglect or abandon their young ones soon after birth. It turns out that there is a difference in the number and distribution of receptors for oxytocin and vasopressin across the limbic brains of these two species, and that each hormone plays a slightly different role in males and females.[18]

If you give a female vole of the prairie-dwelling species oxytocin, the hormone will work like Cupid's arrow and she will become attached to the nearest male that comes her way. Oxytocin exerts its effects by interfering with dopamine reward mechanisms: it binds to receptors in the nucleus accumbens which is one of the reward areas. Female voles of the mountain species have fewer oxytocin receptors in the nucleus accumbens.

In male voles, vasopressin plays a bigger role. In the prairie species it is vasopressin that by binding to receptors in the ventral pallidum – another reward area just below the nucleus accumbens – stimulates pair bonding, aggression towards male rivals and paternal instincts. The higher the number of vasopressin receptors in male prairie voles, the stronger their social attitudes would be.

Voles are one thing, but what about humans? A study looking at a gene associated with the production of vasopressin receptors has reported that men with a particular form of this gene that results in their having few vasopressin receptors in the brain are twice as likely as men with more receptors to stay unmarried or to experience more crises during relationships, with higher risk of divorce.[19] Of course, this is only a correlation and carrying a particular form of a gene is only one of the ingredients contributing to a behavioural tendency.

In sum, it's hard to say with precision, but maybe something was wrong with Madame M's levels of dopamine and oxytocin, because even though she instigated novelty by creating new identities for her husband, she was never carried away by them. Difference, for her, lacked the qualities of the original.

## Choosing a partner

Falling in love with the wrong partner – someone who does not requite our feelings or is not suitable to make us happy – is not a course we would rationally choose, or willingly, for that matter.

Yet, it is something we do. In some cases, we may even do it repeatedly before we find the right soulmate. Without realizing, we systematically follow a pattern of failure. Paradoxically, those who to the objective observer appear blatantly wrong and implausible partners for us, may in fact, for some non-evident reason, appear highly desirable to us, simply fitting a pattern of mismatch.

What makes us fall for the wrong person, or what makes another human being eligible for a union with us, depends on a variety of factors. Some are rooted in childhood.

> They fuck you up, your mum and dad.
> They may not mean to, but they do.
> They fill you with the faults they had
> And add some extra, just for you.

As Philip Larkin accurately observed in his lapidary poem entitled 'This Be The Verse', referring to the inevitable, unintended, powerful influence parents have on us.[20] Indeed, our patterns of approaching, loving and attaching to others are the shadow of ways of loving learnt during childhood, primarily from our parents. Early-life experiences and relationships do affect our adult personality, especially in the realm of intimacy and affection.

These ideas germinated with Freud, but were later extensively explored by the British psychiatrist John Bowlby (1907–91), who wrote: 'When an individual is confident that an attachment figure will be available to him whenever he desires it, that person will be much less prone to either intense or chronic fear than will an individual who for any reason has no such confidence.'[21] For Bowlby such confidence is built during crucial years of infancy, childhood and adolescence.

Years after Bowlby's observations, a significant body of research has confirmed this.

If, on one hand, parents are distant, self-involved and neglectful, the child will consider those attributes acceptable and rewarding and will most likely look for them in his or her partner as an adult. If, on the other hand, children grow up in the presence of warm, gentle, loving and reliable parents, as adults they will most likely develop and appreciate those qualities in others. A mother, or another caregiver, who is responsive to her children's moods and needs is likely to teach them to be loving and seek love. Children learn that when they are in need of help, they can express their needs, and their request for help will be heard. They learn that they are worthy of love and attention and won't be fearful of separation.

Once this dynamic of trust is established early on, children will rely on it throughout their lives and will likely expect it from and create it with the people they meet. So as children we can acquire specific habits, tastes and preferences for relationships, and these will characterize our adult lives, even our romantic lives.[22]

Today, neuroscience is trying to bring these psychological findings to a further level by exploring how early experience is wired on to the brain to steer adult behaviour. In other words, how early parental care gets under our skin.

This fascinating aspect of life and biology has been greatly investigated in animal models, such as rodents, specifically rats and mice, and studies have concentrated in particular on the long-term consequences of the disruption of the maternal–infant relationship. I spent a large portion of my post-doctoral work studying these phenomena in the laboratory of Cornelius Gross at the European Molecular Biology Laboratory.

Part of my job was to observe for several hours a day mother mice looking after their pups. In mice, the crucial window in the pups' development is their first three weeks of life. What they experience during those three weeks shapes their adult life drastically. If you have never done it, it may sound absurd, but you can tell whether a mouse is taking good or bad care of her newborns by watching her. Except for when she needs to eat or drink something, a 'good' mouse mother spends a lot of time with the litter in the nest. She gives them warmth by covering them with her body, assuming a sort of 'blanket' position over them when they all sleep together. She also licks them and grooms them. If one of the pups leaves the nest, she rushes to retrieve it. By contrast, a 'bad' mouse mother is less dedicated to her offspring. She is neglectful of them and spends considerably more time off the nest. When the babies are asleep, she is not very good at covering them with her body and she doesn't care very much for it. She also doesn't bother licking or grooming them. A strong outcome of this difference in mothering style is that the pups raised by a bad mother grow up to be more fearful than those raised by a good mother, as Bowlby would have expected.[23] However, pups born to a bad mother are less fearful as adults if adopted at birth by a more caring mother. Astonishingly, the girls among the group adopted by a good mother go on to acquire her maternal behaviour, displaying a caring attitude towards their own pups, despite their 'bad' genetic origins.

This means that maternal behaviour is transmitted across generations. Parents behave to their children in large part in the same way their own parents behaved towards them.

Philip Larkin conveys this in the second part of his poem:

But they were fucked up in their turn
  By fools in old-style hats and coats,
Who half the time were soppy-stern
  And half at one another's throats.

Man hands on misery to man.
It deepens like a coastal shelf.
Get out as early as you can,
And don't have any kids yourself.

The fact that the pup of a bad mother can acquire caring maternal behaviour from a good mother is evidence pointing to these long-term effects being mediated by early environmental influence. The most intriguing question in this exciting field of research is: what are the molecular mechanisms by which these early environmental effects in childhood are carried on into adulthood? The answer is epigenetic modification. If genetics is the study of how traits are passed on from one generation to the next via the genome, epigenetics is about traits being passed on across generations independently of the genetic information stored in the DNA. It turns out that the quality of maternal care can impart changes in gene expression that modify the adult behavioural characteristics of a mouse and, through the effects of these changes on its own maternal behaviour (in the case of female mice), propagate these changes across generations.

Although at an early stage of discovery, science has even identified some of the specific genes whose expression is modified, and by what molecular epigenetic mechanisms. One of the molecular mechanisms is methylation. This is the addition of a methyl group – a molecule consisting of one carbon atom and three hydrogen atoms, $CH_3$ – to a cytosine base of DNA. Basically, the methyl group acts like a molecular tag that marks the DNA at particular locations. It binds to DNA regions that are responsible for turning on and off the expression of genes.

I have gone to considerable lengths in the quest to verify the ultimate message that the bond between children and their parents builds a consistent model for the establishment of future relationships. Watching those mothers take care of their pups wasn't always fun. I monitored a couple of dozen mice for four hours a day in the dark, meticulously recording in a notebook each and every move they made. Such are the pleasures of science. I remember finding myself at once amused and puzzled by the mice whose maternal behaviour I used to scrutinize. As I stood in front of those cages, I inevitably made fresh comparisons between what I was measuring and what I remembered of my own mother's parenting style, as well as my grandmother's warmth. Was Nonna Lucia warm enough to my mother? How much time did my mother spend hovering over my cradle? Did she ever assume the blanket position? Did she ever lick me, for that matter? In the lab, colleagues and I used to make jokes about those mice and tried to guess the parenting styles of our respective mothers based on our interpersonal relationships and our failures and successes in love.

In addition to the different expectations we had from the relationship, the incompatibility between myself and the Heidelberg guy may well have resided in the ways each of us received love from our own parents, especially our mothers, and the way we exercised that pattern over the years and through other experiences. Bad habits ossify fast, especially if taken up at an early age. Even when it comes to refusing love, or having to struggle for attachment, if we have done it a couple of times, we learn how to do it again.

I admit I was almost frozen with unease by the fact that the way my mother raised me, especially during my early years, must critically influence my choice of partner. It gave me a pretty clear picture of what I was up against when relating to potential partners – and what they were up against when meeting me! Such truth had been handed down to me through concepts in psychoanalysis, as well as Larkin's poem. Neuroscience gave me an additional piece of information. It made me realize that maternal care modifies the expression of some of my genes. This was disquieting, but not totally discouraging. No parent is perfect, but neither is he or she disastrous.

Even though there may be a certain coherence between the kind of person with whom we tend to fall in love and an earlier attachment pattern, we must not think of attachment or parenting styles as shackles that chain people to an immutable sentimental fate. Whether a parent is cold and neglectful or warm and caring is merely the initial impetus in the trajectory of an individual's life. Across the years we can undergo so many changes and accumulate such diverse experience that much of the way we relate to other people is influenced by a lot more than just our parents. As we learnt in chapter 3 in the case of fear, the brain is plastic: its neuronal wiring and the genetic expression underlying it can be actively changed. Epigenetic modification continues even after childhood. Whatever happened in childhood, there is still room for change, development and discovery. Getting tangled up in a pattern is way easier than getting out of it, but the reverse is not impossible. We just need to work towards it, sometimes very hard.

## The love supermarket

My unfortunate experience with the stranger from the cinema confirms that love is blind and that Cupid's lack of judgement can let him make silly, undesirable mistakes on our behalf (or, as Bowlby would explain it, on behalf of our parents). And, when a relationship ends, nobody knows when Cupid will reappear. Nevertheless, for as long as lovers have sought romance, others have tried to meddle, uninvited. Whether they be parents, priests or rabbis, friends or professional matchmakers, third parties have long operated as intermediaries and interfered with the normal course of courtship and flirting, thinking they knew what encouraged lovers to pair off better than the lovers themselves. Most often, they stood against Cupid's arrow so that people would make marital choices that suited social realities. Conventional go-betweens and marriage brokers still exist, but in today's world a new form of matchmaking has come to the fore: online dating.

The online dating industry has become a multi-billion-dollar affair that has continued to prosper even through the ongoing recession, with about twenty-five million individual users around the world having accessed an online dating site at least once in 2011.[24]

Online dating has profoundly changed romance. Most of the time, our encounters are random and unplanned. They occur on the bus, on the street, while we stand in line for coffee, at the supermarket, at a departmental reception, on a plane, or on a boat. We may meet an eligible partner through friends, at weddings or bar mitzvahs, at a dinner party or, as happened to me, in front of the cinema . . . Thus, infatuation with a person begins with only a brief glimpse of who they are. We then gradually discover more and more of their personal qualities, good or bad (or rather, as I emphasized earlier, those qualities we wish to see in them). Nobody can guarantee the success of the relationship, but we embark on it, enjoying the other person for as long as the relationship lasts and is mutually rewarding.

One of the advantages offered by online dating is that, in principle, a new subscriber to a dating site has access to a much larger number of potential lovers than he or she could possibly meet in a more traditional fashion. Compare the dozen or so people you could screen and talk to at a party with the thousands of profiles you can scroll through on your screen. Certainly it would be impossible in practice to meet all users, but a systematic search among them helps the selection, all from the comfort of a desk.

For a fee, dating websites will collect and offer to their subscribers all sorts of basic information about a user: gender and physical attributes as well as self-reported data about personality, background, hobbies, interests and visions of what a relationship should be. All in the form of a refined standard profile and accessible with a couple of clicks. Dating websites ask you to fill out a psychological questionnaire. Once that's done, they pull out matches for you based on compatibility algorithms. On most websites, users can also integrate their basic profiles with unique information through the use of self-descriptions. This is their chance to whet the appetite of those who visit their profiles. To get the feel of such paragraphs, I signed up on one of the dating sites (an action that requires on average a good thirty minutes of your time if you want to answer all the psychological questions) and found that they actually often end up pretty much alike and are written in a repetitive language. Many online daters are not always truthful in reporting their basic information. Of course, in traditional face-to-face dating too, people often tend to strategically offer a slightly better version of themselves to impress the other. However, this deceitful attitude is easier online because of the safe cyber distance. A study of eighty online daters revealed that 81 per cent of them lied about their weight, their height or their age.[25]

Online dating overrides the traditional role of sight in instigating romance.

Courtship is profoundly a physical experience. The standardized profiles, even though incorporating photographs, deprive their subjects of one dimension. Men and women are reduced to two-dimensional profile pages with no movement, glittering glance, or indeed unique smell. Such a system does not remove all element of surprise, possibly unwelcome surprise: a large number of romances begun over the internet end when the prospective lovers meet face to face.

Emotions best travel between us through our bodies. We need the kind of skin reaction offered by a physical encounter, which no computer-based acquaintance can replace. Even the most recent, honest and undoctored pictures can deviate from reality. An attractive photograph may show a pair of finely cut cheekbones, a well-proportioned nose, chiselled lips, even a fit body. But it won't supply the feeling of what it means to be in the presence of those bodily attributes. If, as we have seen, meeting a lover for real sends us on a spectacular trip of fantasy, think how our imagination might travel in the absence of a physical encounter. Well, appropriately, it would go at the speed of the internet.

Too much online dating can cause a desertification of our emotions, to the point that we prefer the two-dimensional profile picture of a body on a screen to a real person at the other side of a table, or in our beds for that matter. People become shopping items, and the dating world a marketplace.[26] There is so much choice available that it is possible to fill a virtual shopping cart, pick and discard at random, and always find a replacement when one of the selected products does not really work out. Thus, rather than educating our senses and emotions to focus on what and who uniquely fulfils our needs and desires – and then work towards achieving that relationship – it becomes easier to quickly consume one product after another. There will always be some other choice. Selecting a potential match online becomes a mechanical, controlled action, comparable to ticking a box in a questionnaire, in stark contrast to the unpredictable and erratic dynamics at work between people in person. Such widespread methods and mentality in love matters corrode the poetry and, ultimately, the trust that we need to build to establish any long-lasting bond.

If online dating in general substitutes calculation for intuition, this is even more the case when biological information is introduced into the business.

An increasing number of innovative dating services, such as Scientificmatch.com, Genepartner.com and Chemistry.com, have integrated their customers' biological information into their selection methods, to match people on the basis of their genetic and chemical profiles. Through the inclusion of this type of data, microscopic fragments of the body enter the picture. These new services have achieved huge popularity – millions of users, at least in America, choosing to consign their romantic fates into the hands of science – and have raised hopes of swift and better success among those seeking their sister soul. Users are persuaded that the help of brain chemistry may be more effective than traditional methods in reversing their repeated failure to find love. But, is it so?

Helen Fisher, one of the first scientists to study love in a brain scanner, has helped build the matching system Chemistry.com. The system was developed around the identification of four main personality types, each reflecting differing levels of two principal neurotransmitters – dopamine and serotonin – and two sex hormones – testosterone and oestrogen.[27]

The personality types are the Explorer, the Builder, the Director and the Negotiator.

When you subscribe to the service, nobody will measure your levels of neurotransmitters and hormones directly, but you will be asked to complete a psychological questionnaire of approximately sixty items that help trace back to them. The questions are based on genetic and neurochemical information linking these four chemicals to personality traits. According to the answers given, you will be attributed a primary and a secondary personality type.

One of the questions is to compare and measure the length of the index and ring fingers of your right hand, when you look at it with its palm up. Why on earth would one want to do that? This again has to do with your mother's influence on your behaviour, which begins in the womb. Chemistry.com looks primarily at the levels of oestrogen and testosterone that filter through the foetal brain. If, as a foetus – male or female – you were exposed to more testosterone, your ring finger will be longer in relation to your index finger. This also reflects higher circulation of testosterone in your body as an adult. A long ring finger and more testosterone means you would be deemed at Chemistry.com to be a Director, a personality type characterized by features such as decisiveness, dominance, directness and self-confidence. Dopamine is connected to a tendency to seek novelty and adventure. Dopamine levels are, therefore, brought into consideration by asking the user how applicable they find such statements as 'I am always looking for new experiences' or 'I find unpredictable situations exhilarating' or 'I do things on the spur of the moment'. If you can strongly relate to these statements, you are labelled as an Explorer. Builders are concrete, cautious, grounded, orderly and with a solid sense of duty. Fisher believes that such properties in a Builder are predominantly orchestrated by serotonin and its influence on the metabolism of hormones and other neurotransmitters. For instance, a Builder's friendliness and tendency to create a family may reside in serotonin's capacity to trigger the release of oxytocin, which, as I explained above, facilitates attachment.

Conversely, a Builder's calm and caution may be in part due to the ability of serotonin to suppress the release of testosterone and dopamine.

Negotiators are intuitive, expressive, pleasing and empathic. They appreciate emotional intimacy and are curious about other human beings. Negotiators have high levels of oestrogen, inherited in the womb from their mother's blood and placenta. Chemistry.com tests the presence of excess oestrogen in Negotiators by checking if their index finger is equal or longer in length than the ring finger. A Negotiator's high oestrogen levels are also assessed on the basis of their enhanced imagination and ability to connect and integrate thoughts and different kinds of information in novel, unexpected ways – in part due to oestrogen's ability to build a high number of nerve connections across distant brain regions within each hemisphere, and between the two hemispheres.

We are left with the question of whether online neuroscience-based matching systems might really be more effective than traditional methods in scouting for an ideal partner. Helen Fisher has helped thousands of romance-seekers find their perfect match.

Patiently, and with a lot of curiosity, I took the test and I can proudly announce that I am a Negotiator-Explorer. Many of the features corresponding to these personality types, as described on the website – such as my high regard for emotional intimacy and the desire to seek new adventures – do actually correspond to some of my dispositions and how I see myself. My ring finger is indeed shorter than my index finger. Yet I am uncomfortable with limiting the totality of who I am to these two attributes. As we saw in chapter 4, conjecturing and identifying 'types' among individuals is not a recent enterprise. Ancient doctors parcelled their population into exemplars of sanguine, choleric, phlegmatic and melancholic tempers. Modern psychology has developed and consistently relied upon inventories of personalities.[28] The desire to understand ourselves, describe our behaviour or intuit that of others is relentless.

It is important to remind ourselves that the four dating types and the matches built with them are not an exact reflection of levels of serotonin, dopamine, oestrogen and testosterone. Being a Negotiator cannot only be the result of an excess of oestrogen and personality types are never the outcome of a single or just a few biological factors. As Fisher acknowledges, 'families' of chemicals and neurotransmitters concoct the types she designed. Behavioural and emotional features arise from a biological architecture that makes them possible, the variation of which confers on individuals personal and unique shadings of those features. Then, as we have seen, chapters in our biographies, environmental circumstances and social and cultural influences play a huge role, too.

The union of two people requires they courageously abandon the safety of their own solitary spaces, to include, make space for and partake of a different – sometimes entirely different – world. It demands that the individuals involved be able to understand and overcome their differences and appreciate the way the other thinks, and imagines life. This is an interesting but entirely insecure journey. Online dating sites that employ scientific information claim that they let users find their ultimate, long-lasting match because a match based on molecular information is more likely to be successful. But even if the psychological compatibility discovered through these methods has scientific rigour, the idea that a long-lasting match may be grounded on the information of a few hormones goes against a basic requirement for the success of a love relationship: that the two individuals *learn* how to love each other and commit to spending a life together, despite their differences. Traits are not immutable. What these dating services are able to offer is the starting basis for a rapport, the appropriate chemistry through which the union has a chance to begin. Sometimes it will last, sometimes not. Because of our hectic life, our increased mobility and the widespread dissolution of traditional modes of courting and socialization in general, online-dating shortcuts to a romantic match may sound more practical, appealing and effective than conventional methods. However, the likelihood of success is not guaranteed to be any higher for a relationship started via a science-based dating platform than for one stemming from a random encounter. Personally, I hope that traditional encounters won't become extinct.

## Coda

It is an interesting circumstance that, in search of inspiration and comfort before a date, I found myself absorbed in Plato's writings, rather than the details of a laboratory experiment. The allegory of the charioteer and the two discordant winged horses echoed the madness of love and the struggle between controlling or yielding to it. But what the allegory ultimately symbolizes is a question which is at the core of this book and the essence of love: how emotion meddles with reason.

Can we employ reason in love, when love is complex, acknowledges no laws, is evanescent and by definition a form of insanity? Can we, and should we, resort to science for matters of the heart?

As an intrinsic and tangible part of our lives and the natural world, love deserves investigation. We are entitled to understand its attributes, build up experience about it and make sense of its unpredictable outcomes the best way we can. Nothing should stop our curiosity to learn about love. Molecules, and scientific empiricism in general, add to the heap of knowledge that is already at our disposal to make sense of it.

However, the amount of scientific data on love is modest compared with what has been said and produced for millennia on the subject of love in the absence of a clear scientific explanation. Because of the dearth of reliable and unequivocal data, I believe that another kind of empiricism, one based on first-hand experience or trial and error, should remain a source as good as, if not more valuable than, any information we may gather from close inspection of a brain in an fMRI scanner.

Certain aspects of love are simply not amenable to scientific investigation. Most studies seeking to dissect romantic love have been limited to mapping its neural anatomy and describing some of its molecular components. Such findings are illustrative of the power of science to reveal the invisible wonder of a phenomenon, but are of little use when we encounter love in our lives. A philosophical or literary work such as a Platonic dialogue or a Shakespearean sonnet can teach us about love's blindness better than can a brain scan or a hormonal test, and will prove more instructive to those in search of tips or desirous to understand the course and excitement of courtship and love. They resonate more loudly and lastingly with anyone looking for experiences with which to *identify*. For instance, what Stendhal called 'crystallization' is a phenomenon we can all grasp without mapping it on the brain. Knowing that gazing at the picture of a beloved dampens the flow of oxygen in brain areas responsible for formulating judgements on the person we are looking at can do little to save us from misattributing qualities to the creature in question.

An ambition arising from the neural and molecular investigations on love has been that of understanding its chemistry in order to exploit it, such as in the case of science-based online dating websites. In general, the employment of science in the search for a sister soul is an attempt to replace the randomness of love with some kind of certainty. It signifies a belief that we can first rigorously choose who would be the best person to fall in love with, and then fall in love with them. But this would be to turn love on its head and dispel its enchantment. In addition, science seems to focus on what it takes to start a love relationship, on how to ignite sparkling romance.

The successful rapprochement of two human beings who aspire to share love depends on an intricate balance of factors that are hard to pin down and orchestrate. On one hand, we have the mark left by our parents, their genetic contribution and their style of upbringing. On the other, we have our own genes and a few unforgiving neurotransmitters that circulate in our bodies. Add the unending and unpredictable everyday experience that moulds our neurons and shuffles our emotions. Then comes social structure and our place in it, the matching of cultural and educational backgrounds as well as personal or recreational interests, let alone political views – I am sure I have left out other important subtle factors here.

I know . . . The totality of these elements makes the alignment of two life-trajectories appear so rare it would make a solar eclipse mundane.

It may be a commonplace observation, but I have come to believe that love simply occurs when two individuals happen to feel a mutual attraction, enjoy each other's company, are keen to embark on adventures together, and are also on the same page and willing to try out the match.

Sadly, it is often the case that when the other person is open to a relationship, we are fearful or, the other way round, when we are ready, they are not – again those patterns . . . And, if someone's heart is not open, there is little we can do to unlock it. There are no flowers, poems or charming surprises that may persuade them to yield. Our persistent attention can definitely help, but if they regard themselves as unworthy of love, even if we tell them they are, they won't believe it until they discover it themselves. This normally has nothing to do with our talents. We may have several respectable qualities on offer, but until our objects of desire are at ease with their own, ours won't make the right impression. They may just prove intimidating. Equally, before going on the prowl, it definitely helps to check first how much we love and consider ourselves worthy of appreciation.

Love is joy's brother. To feel love, it helps to be joyful. And here I mean the joy of a smile as well as the awareness of who one is that will at least give others a reasonably clear picture of the person who desires them. I find it helps to be tenaciously passionate about what one likes and what one detests, enthral an object of desire with enthusiasm and basically show them that being in one's company is the best thing that could ever happen to them. It also helps, through behaviour and actions, to show and reassure the other how sincere we and our feelings are.

Even though the picture may look dim and far from simple, I am not trying to discourage anyone from pursuing or understanding love. I personally prefer to let myself be captured by love in all its uncertainties and forms of expansiveness. We live in a society that incites us to achieve and succeed rather than attach and love. As a consequence, the world seems to reward solitude rather than companionship, and to put into jeopardy the kind of self-effacing attitude and commitment needed to create a trusting relationship. Fear of loving is widespread. Fundamentally, fear of love is fear of risk. We are scared to take chances, make mistakes, be hurt or waste an opportunity. We prefer safety and expect to have guarantees. The use of science to prescribe romance, emotional compatibility and loving relationships that won't fail reinforces our fears and our desire for certainties. It also propagates the idea that we can predict love outcomes. But too much caution and calculation are the wrong approach to love. They won't lead us far.

We would do a tremendous disservice to our own and everyone else's happiness if we saw love as something that has its best destination already set.

In my opinion, what counts most in love is the art of the journey, the fragile enterprise of building trust, day after day. Love is knowledge. It means to create spaces for mutual respect and for the unexpected. It means to evolve both individually and as a pair, with gratitude and responsibility.

Love should also be adventurous. In my experience, it's preferable to take a few bumps here and there rather than have a closed heart. For when love is ripe – even between two people who didn't regard themselves as lovers at first – it doesn't take no for an answer. It falls like a sudden rain when you are under no roof and carrying no umbrella, imposing itself between two beings with the greatest power of persuasion and saying: you don't need a shelter, I am the shelter.

# Epilogue

> No theory of life seemed to him to be of any importance compared with life itself
>
> OSCAR WILDE

I began by questioning whether knowledge about the brain can be of help in understanding ourselves and our emotions in the twenty-first century and I hope that, throughout the pages of this book, I was able to exemplify when neuroscience did shed light on my path, but also the instances where neuroscience just wasn't enough.

When I experience or examine an emotional incident along my trajectory as a man, a friend, a lover, a son or a colleague, the first reservoir of knowledge I consult for explanations and meaning is hardly ever neuroscience. Or, I should say, it's not exclusively neuroscience. I search for and side with the explanation that is most apt for an understanding of what I am feeling, regardless of whether the explanation comes from a scientific experiment, a work of art, a poem, a philosophical theory or even other sources, including my own past experience with a particular emotion.

By no means am I trying to suggest here that neuroscience is inadequate in addressing emotions. In the relatively recent past, the science of the brain has provided us with fresh accounts of how we emote, some of which may well resonate with us. It's hard not to be fascinated by Damasio's theory of emotions and somatic marker hypothesis. The fact that emotion guides reasoning overturns centuries of mistaken assumptions about our rationality and the way we face choices. That our emotional experience writes itself somehow in our bodies, in our neurons, to guide our instinct and intuition, and that we may have discovered where in the brain this inscription occurs is an irresistible notion. Equally, the discovery of the plasticity of the brain is of great relevance if we think of its meaning and importance in, for instance, overriding unwanted patterns of fear, or even honing our approach to love. There is endless wonder in the images of neuroscience. Yet they do not cover the entire breadth of an emotion.

When I describe an emotion in scientific terms, I always wonder: is what I am saying correct? Am I doing my emotions justice? Am I doing science justice, for that matter? When listing brain regions, nerves or competing chemical actions, I do marvel at how something as complex and at the same time ephemeral as emotions can be confidently translated into discrete detailed models, but I always bear in mind that there is a distance between what such details describe and what I feel.

This brings me to another reflection. Most of what we first learn about life, the nature of human beings and their emotions emerges from life itself, from our personal vicissitudes. My own subjective account of emotions is free of the constraints of science. There are no borders within which to fall, no molecular nomenclature to respect. It is simply what I feel: a rich, intimate speech that the language of science cannot and will not – well, at least not in my lifetime, I believe – replace.

Such direct and immediate appreciation of emotions is a plane of knowledge at the heart of everyone's existence. It is a speech that belongs to us alone. The objective, third-person, detailed accounts of what we suppose is taking place in the brain as we speak, cry, laugh, feel guilty, miss or love somebody can be valuable, fascinating additions, but are sometimes only minor footnotes.

So, knowledge of the detailed neural subtext of brain tissues, neurons, stretches of DNA and molecular fluctuations does not always contribute to composing the daily script of our emotional lives. To fill those gaps and cover the distance that separates us from understanding our emotions, we are entitled to take all kinds of shortcuts. Many different roads lead us in the direction of *Know Thyself.*

As citizens of life and consumers of knowledge in a time when science dominates the public discourse, we can learn how to skilfully and harmoniously integrate science teachings, art, poetry, philosophy as well as our own observations as human beings. Throughout my life, I simply haven't been able to disjoin these various ways of looking at the world – they have belonged to the same library shelf. And that's because no view is on its own sufficient or satisfactory. There is absolutely no reason to live by only one set of ideas and not be curious about or open to others. All approaches will always leave questions unanswered. There will always be more to discover.

Take a look at the picture below and ask yourself what you see.[1]

At first, you might notice a duck's beak, then a rabbit's pair

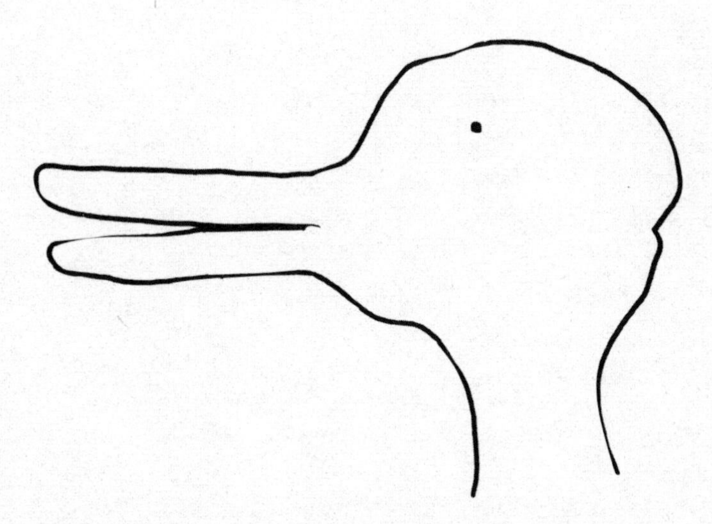

of ears, or maybe the other way around. What you see are not just two different animals. Each can be taken to represent a coherent system of looking at the world. Say one is science, the other is the arts and the humanities – you choose which is which. These two world-views intersect at once harmoniously and discordantly. There may be people who only see the duck, others only see the rabbit. But most of us, at least if made aware of the double aspect of the picture, should be able to easily switch from one version to the other. What we must remember is that truths are fleeting. One day the rabbit may disappear or even gulp down the duck. Until we can favour one version and drop the other, the two interpretations of the same phenomenon will coexist, and neither is more or less meaningful or valid as explanation than the other. Rather than bringing either enchantment or disenchantment, each vision complements the other and shapes a thorough world-view.

While neuroscience explains emotions through figures and measurements, predicting causes and outcomes, how we understand emotions will always rely on more than just science. It is possible to be at once scientific and lyrical when we attempt to understand ourselves and how we feel.

# Acknowledgements

Carrie Kania, my literary agent, earns the first mention. I still remember the evening she sat down with me in a London restaurant to encourage me to write this book. I thank her heartily for her warm generosity and friendship and I salute her sharp wit and unflagging enthusiasm about books and ideas. I also thank Patrick Walsh for his precious advice at the dawn of the project and Alexandra McNicoll, Henna Silennoinen and Jake Smith-Bosanquet at Conville and Walsh for their kindness and invaluable work.

I am enormously grateful to my editors Doug Young, from Transworld Books, and Allison Lorentzen, from Penguin, for eagerly embracing the idea of this book and for all their support and advice.

Noga Arikha, Stephanie Brancaforte, Allen Frances, Helga Nowotny, Steven Rose and Donna Stonecipher provided me with illuminating comments and criticism on manuscript drafts. I am indebted to them for their precious time and insights.

During its long period of gestation, this book benefited from the help of two institutions and their libraries: the Berlin Institute for Cultural Inquiry and the Wissenschaftskolleg zu Berlin.

There are several mentors who throughout the years have been central to my intellectual and creative path. They have all turned into splendid friends and have my deep respect and gratitude. I owe to Halldór Stefánsson, from the Science and Society programme of the European Molecular Biology Laboratory (EMBL), the initial impetus to frame science within a larger context. Helga Nowotny has generously offered friendship, acute advice, as well as continuous and encouraging guidance on many of my choices, even the most adventurous. Cornelius Gross, my post-doctoral supervisor at EMBL, opened the doors of his laboratory to me and was always available for broad engaging discussions about neuroscience.

Nikolas Rose and Ilina Singh at the BIOS Centre of the London School of Economics took me under their wings in an unfamiliar world. Suzanne Anker from the New York School of Visual Arts has been a great chaperon in the visual art world.

Many thanks also to my colleagues from the European Neuroscience and Society Network, especially Linsey McGoey and Scott Vrecko and all the alumni of the transdisciplinary Neuroschools, for all the shared efforts and great times spent building a forum for innovative discussions across neuroscience, the social sciences and the humanities.

Many thanks to the one and only Ben Crystal for his friendship and endless chats about theatre, Shakespeare and love; to Donna Stonecipher for her help with poetry metrics. Chapter 2 is dedicated to Alexander Polzin whom I thank for his input on Caravaggio. I also thank my friend Sabin Tambrea for having inspired sonnets and for the many fascinating theatre evenings at the Berliner Ensemble.

Café Bondi in Berlin Mitte provided all the daily caffeine needed to start my writing in the morning.

There are many friends with whom I have traded tales on emotions and who, near or far, have provided great company and encouragement as I wrote the manuscript: Stephanie Brancaforte, Dominique Caillat, Stephen Cave, Rose-Anne Clermont, Elena Conti, Zoran Cvetkovic, Patrizia D'Alessio, Larry Dreyfus, Amos Elkana, Allen Frances, Valentina Gagliano, Frank Gillette, Marco Giugliano, Manueal Heider de Jahnsen, Christoph Heil, Christine Hill, Stephanie Jaksch, Carlos Kraus, David Krippendorf, Babette Kulik, Luisa Lo Iacono, Sharmaine Lovegrove, Donna Manning, Jimmy Nilsson and Ilaria Cicchetti-Nilsson, Petr Nosek, Alan Oliver, Moritz Peill-Meininghaus, Elisabetta Pian, Marcello

Simonetta, Sabin Tambrea, Anne-Cécile Trillat, Simon Van Booy, Candace Vogler, Mathew Westcott, Katharina Wiedemann, Bonnie Wong.

Silvia Curado was especially there from across the pond for emergency Skype conversations when there was a particular emotional incident, sad or joyful, to figure out.

Noga Arikha was an irreplaceable source of deep, tenacious, truthful exchange on the many beauties and oddities of life.

Roberto and Massimiliano showed up when I had just started to write this book and their joyful company helped ferry it to the right shore.

My loving gratitude goes to Avi Lifschitz who, without knowing it, pointed to what really mattered.

Enza Ragusa has been a rock upon which I have stood since I was just a kid of nine. I thank her for her truthful friendship and unconditional support.

My parents, Giuseppe and Salvina Frazzetto, as well as my sister Antonella, are the recipients of my warmest gratitude for their unwavering trust.

I wish my two smart, sweet, irreplaceable young nieces, Alice and Eva, a life ripe with many great and unforgettable emotional adventures.

Yehuda Elkana partook of this project since its germination, but unfortunately did not live to see it in print. He was a dear and loyal friend, an endless source of strength, joy and wisdom. He is terribly missed and this book is dedicated to his memory, with all my heart.

# Notes and References

## Prologue

1 The episode refers to a passage in Plato's *Phaedo*.

2 Weber, M. (ed. D. Owen and T. B. Strong), *The Vocation Lectures*, Hackett Publishing Company, 2004.

## Chapter 1

1 Darwin, C., *The Expression of the Emotions in Man and Animals* (originally published 1872), in Wilson, E. O. (ed.), *From So Simple a Beginning: The Four Great Books of Charles Darwin*, Norton, 2006. Some of Darwin's theories on behaviour and emotions also appeared in his earlier notebooks.

2 Prodger, P., *Darwin's Camera: Art and Photography in the Theory of Evolution*, Oxford University Press, 2009.

3 For an eloquent and thorough contemporary insight into emotions as evolved neural circuits and their study in lower animals, see LeDoux, Joseph, 'Rethinking the emotional brain', *Neuron*, 73 (2012), 653–76.

4 This distinction, in somewhat different terms, had already been made at the end of the nineteenth century by the American psychologist William James, whose work I mention in chapter 3. The neuroscientist Antonio Damasio has greatly refined and extended this through his own work. See Damasio, A. R., *Descartes' Error: Emotion, Reason and the Human Brain*, Penguin, 2005, and *The Feeling of What Happens*, Harcourt Brace & Co., 2000.

5 Between the 1960s and 1980s, the psychologist Paul Ekman collected an extensive set of pictures and data from remote places, such as Papua New Guinea, to confirm Darwin's theory and to map the expression on several face muscles. See Ekman, P., *Emotions Revealed*, Henry Holt and Company, 2003.

6 Smith, C. U. M., 'The triune brain in antiquity: Plato, Aristotle, Erasistratus', *Journal of the History of the Neurosciences*, 19 (2010), 1–14.

7 Freud, S., *New Introductory Lectures on Psychoanalysis*, W. W. Norton and Company, 1933.

8 For a thorough account of a triune neuropsychology, see Paul McLean's seminal book *The Triune Brain*, Plenum Press, 1990.

9 For an interesting and detailed explanation of the evolution of the neocortex, see Rakic, P., 'Evolution of the neocortex: Perspective from developmental biology', *Nature Reviews Neuroscience*, 10 (2010), 724–35.

10 It is important to mention another, complementary way of classifying our emotional and thinking selves. The brilliant psychologist Daniel Kahneman distinguishes between two main systems at our disposal for processing facts and knowledge and for making decisions. System One is fast, intuitive, unconscious and irrational. System Two is slow, logical, conscious and rational. System One makes us take decisions in a fraction of a second, while System Two is more critical and formulates judgements after careful consideration. The latter is evolutionarily younger than the former and it also consumes more energy. System One and System Two may even be loosely compared to Freud's id and ego. See Kahneman, D., *Thinking Fast and Slow*, Farrar, Straus & Giroux, 2011. For an explanation of the differences between Kahneman and Freud's thinking, see Dyson, F., 'How to dispel your illusions', *New York Review of Books*, 22 December 2011.

11 For a full description of the accident, see Damasio, *Descartes' Error: Emotion, Reason and the Human Brain.*

12 For the excellent report on the analysis of Phineas Gage's skull, see Damasio, H., *et al.*, 'The return of Phineas Gage: Clues about the brain from the skull of a famous patient', *Science*, 264 (1994), 1102–5; for the original report by Gage's doctor, see Harlow, J., *Publications of the Massachusetts Medical Society*, 2 (1868), 327.

13 While I was writing this chapter a man in Brazil had an accident similar to that of Phineas Gage. This new patient was not a miner, but simply had his brain perforated by a rod that entered from above. This new case might finally give Gage a rest, but, of course, it will take time until the behavioural consequences of the accident become prominent and can be studied with scientific rigour. See MacKinnon, Eli, 'Eduardo Leite dubbed modern-day Phineas Gage after pole pierces his brain', HuffPost Science website, 22 August 2012.

14 Bechara, A., Damasio, H., Tranel, D., and Damasio, A. R., 'Deciding advantageously before knowing the advantageous strategy', *Science*, 275 (1997), 1293–5.

15 Interview by David Brooks on FORA.tv: http://fora.tv/2009/07/04/Antonio_Damasio_This_Time_With_Feeling

16 It is important to note that the prefrontal cortex is anatomically and functionally heterogeneous, so that different 'sectors' in it, if damaged, have variegated consequences. It is also useful to bear in mind that the lesions observed in Gage and other patients encompass large sections of the prefrontal cortex. When scientists examine the lesions using modern visualization techniques, they do their best to localize them and define their perimeter as closely as possible in order to assign shadings of behaviour to several sub-territories. For a good review of the prefrontal cortex and violent individuals, see Yang, Y., and Raine, A., 'Prefrontal structural and functional brain imaging findings in antisocial, violent, and psychopathic individuals: A meta-analysis', *Psychiatry Research*, 174 (2009), 81–8; for a review of the role of the PFC in social cognition, see Amodio, D. M., and Frith, C. D., 'Meeting of minds: The medial frontal cortex and social cognition', *Nature Reviews Neuroscience*, 7 (2006), 268–77.

17 Blair, R. J. R., and Cipolotti, L., 'Impaired social response reversal. A case of "acquired sociopathy"', *Brain*, 123 (2000), 1122–41.

18 Interestingly, patients who had the frontal regions of their brains compromised in infancy or childhood had lifelong and more severe behavioural changes in comparison with those documented for Phineas Gage or patients such as Elliot or Jay, whose brains were injured in adulthood. The greater severity of their insensitivity to moral and social rules and their inability to learn social cues may mean that the regions of the brain that were injured also play a role in the acquisition of social knowledge. For a report on two cases of patients with prefrontal cortex lesions that occurred in infancy, see Anderson, S. W., Bechara, A., Damasio, H., Tranel, D., and Damasio, A. R., 'Impairment of social and moral behavior related to early damage in human prefrontal cortex', *Nature Reviews Neuroscience*, 2 (1999), 1032–7.

19 Raine, A., Meloy, J. R., Bihrle, S., Stoddard, J., LaCasse, L., and Buchsbaum, M. S., 'Reduced prefrontal and increased subcortical brain functioning assessed using positron emission tomography in predatory and affective murderers', *Behavioural Sciences and the Law*, 16 (1998), 319–32; Raine, A., Buchsbaum, M., and LaCasse, L., 'Brain abnormalities in murderers indicated by positron emission tomography', *Biological Psychiatry*, 42 (1997), 495–508.

20 Davidson, R. J., Putnam, K. M., and Larson, C. L., 'Dysfunction in the neural circuitry of emotion regulation – a possible prelude to violence', *Science*, 289 (2000), 591–4.

21 An experiment conducted in 1999 by Dr Shiva and his colleagues elegantly demonstrated the battle between emotions and reason as

controlled by the PFC by showing how the PFC can manage only a few cognitive tasks at a time. Participants were asked to use their short-term memory to remember a number and then were asked to choose between a bowl of fruit salad and a chocolate cake. Those who memorized a seven-digit number were unable to resist the temptation of choosing chocolate cake over the fruit salad. On the contrary, participants who only had to memorize a one-digit number were able to use their PFC to exercise will power and opt for the healthy food. With a lower cognitive load, you can have more will power and resist temptations. The PFC also plays a role in working memory, which is about storing information for further use and manipulation. I discuss this in chapter 6.

22 Brunner, H. G., *et al.*, 'X-linked borderline mental retardation with prominent behavioural disturbance: Phenotype, genetic localization, and evidence for disturbed monoamine metabolism', *American Journal of Human Genetics*, 52 (1993), 1032–9.

23 Brunner, H. G., *et al.*, 'Abnormal behaviour associated with a point mutation in the structural gene for monoamine oxidase A', *Science*, 262 (1993), 578–80.

24 The reason these men did not produce MAOA is that they had a mutation in the MAOA gene on the X chromosome that resulted in a premature halt in the production of the enzyme.

25 Sabol, S., *et al.*, 'A functional polymorphism in the monoamine oxidase A gene promoter', *Human Genetics*, 103 (1998), 273–9.

26 Cases, O., Seif, I., Grimsby, J., *et al.*, 'Aggressive behavior and altered amounts of brain serotonin and norepinephrine in mice lacking MAOA', *Science*, 268 (1995), 1763–6. You may have noticed some gender bias here, in that I have only mentioned MAOA-related violent behaviour in men. It isn't that women aren't aggressive or can't feel rage or that they don't commit crimes. Since the MAOA gene lies in the X chromosome, it is only from his mother that a man can inherit the low-producing version of the gene. And whereas in his sister (who as a female inherits an X chromosome from each parent) the low-producing form of the gene in their mother's X chromosome can be compensated for by a high-producing copy of the gene in the X chromosome coming from her dad, in the man's case the paternal Y chromosome can't help. That is why behavioural alterations due to defects in the MAOA gene are more evident in males than in females. Of course, any other gene playing a role in violence will have a greater effect in men than in women if it is also X-linked.

27 Rose, S., and Rose, H., *Alas Poor Darwin: Arguments against Evolutionary Psychology*, Random House, 2000.

28 By way of clarification, it should be said that, in the case of the Dutch kindred, the mutation in the sequence MAOA gene was such that no enzyme could be produced. In such cases the mutation had a direct effect on the onset of criminality, but the long or short variation of MAOA alone is not a sufficient cause for criminality.

29 Rakersting, A., Kroker, K., Horstmann, J., *et al.*, 'Association of MAO-A variant with complicated grief in major depression', *Neuropsychobiology*, 56 (2008), 191–6; Frydman, C., Camerer, C., Bossaerts, P., and Rangel, A., 'MAOA-L carriers are better at making optimal financial decisions under risk', *Proceedings of the Royal Society*, 278 (2010), 2053–9.

30 Caspi, A., *et al.*, 'Role of genotype in the cycle of violence in maltreated children', *Science*, 297 (2002), 851–4.

31 Frazzetto, G., *et al.*, 'Early trauma and increased risk for physical aggression during adulthood: The moderating role of MAOA genotype', *PLOS One*, 5, Issue 2 (2007), e486; Widom, C. S., and Brzustowicz, L. M., 'MAOA and the "cycle of violence": Childhood abuse and neglect, MAOA genotype, and risk for violent and antisocial behaviour', *Biological Psychiatry*, 60 (2006), 684–9; Kim-Cohen, J., *et al.*, 'MAOA, maltreatment, and gene–environment interaction predicting children's mental health: New evidence and a meta-analysis', *Molecular Psychiatry*, 11 (2006), 903–13.

32 Gautam, N., 'What's on Jim Fallon's Mind?', *Wall Street Journal*, 30 November 2009. You can also hear Dr Jim Fallon talk about his story in his own TED Talk: http://www.youtube.com/watch?v=u2V0vOFexY4

33 It has been found that the low-activity variant of MAOA predicted volume reduction in limbic areas, hyper-responsive amygdalas and diminished reactivity in regulatory prefrontal regions: Meyer-Lindenberg, A., Buckholtz, J. W., Kolachana, B., *et al.*, 'Neural mechanisms of genetic risk for impulsivity and violence in humans', *Proceedings of the National Academy of Sciences*, 103 (2006), 6269–74.

34 Feresin, E., 'Lighter sentence for murderer with "bad genes"', *Nature*, 30 October 2009.

35 Ibid.

36 For an informative summary of the use of neuroscience evidence in UK courts, see 'Brain Waves 4: Neuroscience and the law', a report published by the Royal Society, London, December 2011.

37 Aspinwall, L. G., Brown, T. R., and Tabery, J., 'The double-edged sword: Does biomechanism increase or decrease judges' sentencing of psychopaths?', *Science*, 337 (2012), 846–9.

38 Eagleman, D., *Incognito: The Secret Lives of the Brain*, Pantheon Books, 2011.

39 Gopnik, Adam, 'One more massacre', *New Yorker*, 20 July 2012.

40 Associated Press, 31 August 2012.

41 Barron, J., 'Nation reels after gunman massacres 20 children at school in Connecticut', *New York Times*, 14 December 2012.

42 Hagan, C., 'Geneticists studying Connecticut shooter's DNA', CNN online, 28 December 2012.

43 In 2002 President George W. Bush launched the new Freedom Commission on Mental Health to monitor the delivery system of the US mental health service. The plan included mental health screening for the fifty-two million students and six million teachers at educational institutions and appropriate treatment intervention, including drug treatments. Later the plan was dismissed because of conflicts of interest among the politicians who proposed it: they were board members of many of the pharmaceutical companies which would have founded the programme; Lenzer, J., 'Bush plans to screen whole US population for mental illness', *British Medical Journal*, 328 (2004), 1458.

44 For a critical appraisal of the study of dangerous brains, see Schleim, S., 'Brains in context in the neurolaw debate: The examples of free will and "dangerous" brains', *International Journal for Law and Psychiatry*, 35 (2012), 104–11.

45 'Who calls the shots?', editorial in *Nature* after the massacre by James Holmes in Aurora, Colorado, USA, *Nature*, 488 (2012), 129.

46 Widdicombe, Lizzie, 'Shots', *New Yorker*, 3 September 2012.

47 Chang, P. P., Ford, D. E., Meoni, L. A., *et al.*, 'Anger in young men and subsequent premature cardiovascular disease', *Archives of Internal Medicine*, 162 (2002), 901–6.

48 Tafrate, R. C., Kassinove, H., and Dundin, L., 'Anger episodes in high- and low-trait anger community adults', *Journal of Clinical Psychology*, 58 (2002), 1573–90.

49 Eagleman, *Incognito: The Secret Lives of the Brain.*

50 Seneca citations are taken from: John M. Cooper and J. F. Procopé (eds), 'On Anger', in *Seneca: Moral and Political Essays*, Cambridge University Press, 1995.

51 Rose, S., *Lifelines: Life Beyond the Gene*, Oxford University Press, 2003.

**Chapter 2**

1 Freud, S., *The Interpretation of Dreams* (originally published by Macmillan Company, 1913), Forgotten Books, 2012.

2 Ibid. p. 88.

3 Ibid. pp. 80–1.

4 Darwin, C., *The Expression of the Emotions in Man and Animals* (originally published 1872), in Wilson, E. O. (ed.), *From So Simple a Beginning: The Four Great Books of Charles Darwin*, Norton, 2006, p. 1415.

5 This scenario is studied in Zeelenberg, M., and Breugelmans, S. M., 'The role of interpersonal harm in distinguishing regret from guilt', *Emotion*, 8 (2008), 589–96.

6 For an in-depth psychological analysis of guilt and shame, see Tangney, June Price, and Dearing, Ronda L., *Shame and Guilt*, Guilford Press, 2000.

7 However, this distinction is not without exceptions. Often, moral transgressions that cause guilt may not escape the notice of others. If you commit a crime, people you know are going to find out and your name will most probably end up in some newspaper. On the other hand, shame can be solitary. We can be ashamed about something and yet keep doing it secretly; Tangney and Dearing, *Shame and Guilt*.

8 Blushing is also a sign of embarrassment. Embarrassment is similar to shame and guilt in that it is an emotion that concerns the self. However, in contrast, it is fleeting and is also less serious or morally problematic. It is a relatively superficial feeling that arises from, say, an accident for which we feel responsible, but that has no long-lasting consequences and that does not devalue our whole persona. Eisenberg, N., 'Emotion, regulation and moral development', *Annual Review of Psychology*, 51 (2000), 665–97.

9 For a discussion of visceral and moral disgust, see Jones, D., 'The depths of disgust', *Nature*, 447 (2007), 768–71.

10 Moll, J., *et al.*, 'Human fronto-mesolimbic networks guide decisions about charitable donation', *Proceedings of the National Academy of Sciences*, 103 (2006), 15623–8.

11 Interestingly, a subsequent similar study demonstrated that the psychological impact of washing one's hands extends beyond the moral domain and 'washes away' the typical conflict we feel each time we choose something over something else. For instance, choosing whether to go to Rome or to Paris on holiday causes a conflict that while not dramatic is still unpleasant. Both options are valuable. Normally, in order to avoid this conflict, we tend to justify our choice by finding reasons why the preferred choice is far more attractive than the option we rejected. The study showed that washing one's hands after taking such a decision reduces this typical reaction. It removes traces of past decisions and makes the rejected option less unattractive. In other

words, by reducing the participants' need to justify their option, the act of physically cleaning the hands served the purpose of preventing regret; Lee, S. W. S., and Scharz, N., 'Washing away post-decisional dissonance', *Science*, 328 (2010), 709.

12 Atik, A., *How It Was: A Memoir of Samuel Beckett*, Faber and Faber, 2001.

13 Escobedo, J. R., and Adolphs, R., 'Becoming a better person: Temporal remoteness biases autobiographical memories for moral events', *Emotion*, 10 (2010), 511–18.

14 The dilemma of the drowning child is adapted from an example given by Joshua Greene at a conference on the science of morality: http://www.edge.org/3rd_culture/morality10/morality.greene.html; and from Unger, P., *Living High and Letting Die: Our Illusion of Innocence*, Oxford University Press, 1996.

15 Greene, J., 'From neural "is" to moral "ought": What are the moral implications of neuroscientific moral psychology?', *Nature Reviews Neuroscience*, 4 (2003), 847–50. The original study by Joshua Greene and colleagues probing the 'emotional' networks underlying moral judgement is Greene, J. D., Sommerville, R. B., Nystrom, L. E., Darley, J. M., and Cohen, J. D., 'An fMRI investigation of emotional engagement in moral judgment', *Science*, 293 (2001), 2105–8.

16 Basile, B., Mancini, F., Macaluso, E., Caltagirone, C., Frackowiak, R. S., and Bozzali, M., 'Deontological and altruistic guilt: Evidence for distinct neurobiological substrates', *Human Brain Mapping*, 32 (2011), 229–39; Moll, J., Oliveira-Souza, R., Garrido, G. J., Bramati, I. E., Caparelli-Daquer, E. M., Paiva, M., Zahn, R., and Grafman, J., 'The Self as a moral agent: Linking the neural bases of social agency and moral sensitivity', *Social Neuroscience*, 2 (2007), 336–52; Kedia, G., Berthoz, S., Wessa, M., Hilton, D., and Martinot, J. L., 'An agent harms a victim: A functional magnetic resonance imaging study on specific moral emotions', *Journal of Cognitive Neuroscience*, 20 (2008), 1788–98; Takahashi, H., Yahata, N., Koeda, M., Matsuda, T., Asai, K., and Okubo, Y., 'Brain activation associated with evaluative processes of guilt and embarrassment: An fMRI study', *Neuroimage*, 23 (2004), 967–74.

17 Wagner, U., N'Diaye, K., Ethofer, T., Vuilleumier, P., 'Guilt-specific processing in the prefrontal cortex', *Cerebral Cortex*, 21 (2011), 2461–70. I am grateful to Dr Ullrich Wagner for his friendly availability to discuss his work on guilt and moral emotions.

18 Fendez, M. F., 'The neurobiology of moral behavior: Review and neuropsychiatric implications', *CNS Spectre*, 14 (2009), 608–20.

19 A growing field of research explores the body's reactions to looking at art and the empathic engagement of spectators with the feelings

portrayed in a work of art. This research is intended to unravel some of the basic neural mechanisms involved when viewing art. I write about empathy in chapter 5. For a magnificent and comprehensive study of the neurobiology of aesthetic appreciation, see Kandel, E., *The Age of Insight*, Random House, 2012.

20 My main sources for the life of Caravaggio are the excellent biographies by Andrew Graham-Dixon: *Caravaggio: A Life Sacred and Profane*, Penguin, 2010, and Francine Prose: *Caravaggio, Painter of Miracles*, Harper Perennial, 2010.

21 Graham-Dixon, *Caravaggio: A Life Sacred and Profane*, p. 333.

22 Harris, J. C., 'Caravaggio's Narcissus', *American Journal of Psychiatry*, 67 (2010), 1109.

23 Graham-Dixon, *Caravaggio: A Life Sacred and Profane*, p. 333.

24 For an overview of the principles of fMRI, see Logothetis, N. K., 'What we can do and what we cannot do with fMRI', *Nature*, 453 (2008), 869–78; see also Fitzpatrick, S.,'Functional brain imaging. Neuro-turn or wrong Turn?', in Littlefield, M., and Johnson, J. M. (eds), *The Neuroscientific Turn: Transdisciplinarity in the Age of the Brain*, University of Michigan Press, 2012.

25 Jueptner, M., and Weiller, C., 'Review: Does measurement of regional cerebral blood flow reflect synaptic activity? Implications for PET and fMRI', *Neuroimage*, 2 (1995), 148–56.

26 Logothetis, 'What we can do and what we cannot do with fMRI' (supplementary material) and Pauling, L., and Coryell, C., 'The magnetic properties and structure of hemoglobin', *Proceedings of the National Academy of Sciences*, 22 (1936), 210–16.

27 McCabe, D. P., and Castel, A. D., 'Seeing is believing: The effect of brain images on judgments of scientific reasoning', *Cognition*, 107 (2008), 343–52.

28 For an excellent perspective on how brain scans have become iconic images in the public domain, see Dumit, J., *Picturing Personhood: Brain Scans and Biomedical Identity*, Princeton University Press, 2003.

29 As reviewed in Fitzpatrick, 'Functional brain imaging. Neuro-turn or wrong turn?'.

30 Logothetis, 'What we can do and what we cannot do with fMRI' (supplementary material).

31 Website for the IgNobel Prizes: www.improbable.com

32 The study is Bennett, C. M., Baird, A. A., Miller, M. B., and Wolford, G. L., 'Neural correlates of interspecies perspective taking in the

post-mortem Atlantic salmon: An argument for proper multiple comparisons correction', *Journal of Serendipitous and Unexpected Results*, 1 (2010), 1–5.

33 For another excellent critical discussion of the statistical calculations behind fMRI data on emotions, see Vul, E., Harris, C., Winkielman, P., and Pashler, H., 'Puzzlingly high correlations in fMRI studies of emotion, personality, and social cognition', *Perspectives on Psychological Science*, 4 (2009), 274–90. This article refers to another problematic aspect of statistical analysis in fMRI: the fact that when researchers calculate correlations between fMRI data and personality traits or measures of emotion, they often make a separate correlation for each unit (voxel) of the area of the brain involved. They then report results from those voxels that go beyond certain significant value thresholds, thus possibly inflating the correlations.

34 Darwin, C., *The Expression of the Emotions in Man and Animals* (originally published 1872), in Wilson, E. O. (ed.), *From So Simple a Beginning: The Four Great Books of Charles Darwin*, Norton, 2006, p. 1414.

35 The correspondence between Freud's theory of the mind and the brain's physical map has been put forward by Mark Solms, a pioneer in the field of neuropsychoanalysis, a discipline that seeks to confirm theories of psychoanalysis through the methods of neuroscience. For a review, see Solms, M., 'Freud returns', *Scientific American*, May 2004, 83–8.

36 Shamay-Tsoory, S. G., Tibi-Elhanamy, Y., and Aharon-Petrez, J., 'The green-eyed monster and malicious joy: The neuroanatomical bases of envy and gloating (Schadenfreude)', *Brain*, 130 (2007), 1663–78; Takahashi, H., Kato, M., Matsuura, M., Mobbs, D., Suhara, T., and Okubo, Y., 'When your gain is my pain and your pain is my gain: Neural correlates of envy and Schadenfreude', *Science*, 323 (2009), 937–9.

37 For instance, Marc Hauser argues that we are equipped with a universal 'moral grammar' engraved in our biological make-up that makes us formulate moral judgements intuitively. See Hauser, M. D., *Moral Minds: How Nature Designed Our Universal Sense of Right and Wrong*, Ecco/HarperCollins, 2006. It is important to point out, however, that a university investigation concluded in 2010 found Hauser responsible for scientific misconduct, casting doubt on some of his findings. See Johnson, Carolyn Y., 'Ex-Harvard scientist fabricated, manipulated data, report says', *Boston Globe*, 5 September 2012.

38 For a review of the challenges of locating emotions in the brain, see Hamann, S., 'Mapping discrete and dimensional emotions onto the brain: Controversies and consensus', *Trends in Cognitive Sciences*, 16 (2012), 458–66.

39 In the rest of this book I am going to cite several studies involving brain-imaging techniques which have been useful in exploring the neural geography of various aspects of mental life. In each case I will highlight what they have revealed about emotions, but all the technical limitations I have illustrated in this chapter are to be taken into account.

40 Abad, H., *Recipes for Sad Women* (trans. Anne McLean), Pushkin Press, 2012.

41 Wilde, O., *The Picture of Dorian Gray*, Penguin, ebook, 2006.

**Chapter 3**

1 Auden, W. H., *The Age of Anxiety* (originally published 1947), Princeton University Press, 2011, p. 3.

2 Ibid.

3 Ibid. As annotated in Introduction, p. xiii.

4 Campbell, D., 'Recession causes surge in mental health problems', *Guardian*, 1 April 2010.

5 As stated on the NHS fact sheet on generalized anxiety disorder: http://www.nhs.uk/conditions/anxiety/Pages/Introduction.aspx

6 As published on the statistics page for anxiety disorders of the National Institute of Mental Health and based on: Kessler, R. C., Chiu, W. T., Demler, O., *et al.*, 'Prevalence, severity, and comorbidity of twelve-month DSM-IV disorders in the National Comorbidity Survey Replication (NCS-R)', *Archives of General Psychiatry*, 62 (2005), 617–27; http://www.nimh.nih.gov/statistics/1ANYANX_ADULT.shtml

7 Helm, Toby, 'Victims of recession to get free therapy', *Guardian*, 8 March 2009.

8 Collier, R., 'Recession stresses mental health', *Canadian Medical Association Journal*, 181 (2009), 3–4; http://www.nhs.uk/conditions/Anxiety/ Pages/; Smith, K., 'Trillion-dollar brain drain', *Nature*, 478 (2011), 15.

9 Nesse, R., 'Proximate and evolutionary studies of anxiety, stress and depression: Synergy at the interface', *Neuroscience and Biobehavioral Reviews*, 23 (1999), 895–903.

10 Darwin, C., *The Expression of the Emotions in Man and Animals* (originally published 1872), in Wilson, E. O. (ed.), *From So Simple a Beginning: The Four Great Books of Charles Darwin*, Norton, 2006, p. 1432.

11 Ibid. p. 1415.

12 James, W., 'What is an emotion?', *Mind*, 9 (1884), 188–205; in the excerpt I cite, James actually uses the word 'feeling' to describe what we today would call emotions: 'the feeling neither of quickened heartbeats', etc. This is probably due to the usage of the word 'feeling' at the end of the nineteenth century, but James's idea is that bodily changes (emotions) inform our awareness of feelings.

13 In 2011, with choreographer and theatre director Sommer Ulrickson, I created a theatrical performance about anxiety entitled *Fear in Search of a Reason* (echoing Pirandello's *Six Characters in Search of an Author*) at the Institute for Cultural Inquiry in Berlin.

14 Freud, S., *Introductory Lectures on Psycho-analysis (Part III),* Vol. XVI (1917), *The Standard Edition of the Complete Psychological Works of Sigmund Freud.* Lecture XXV: Anxiety, pp. 393.

15 Beard, G. M. (with Rockwell, A. D.), 'Nervous exhaustion (neurasthenia)', Chapter I in *A Practical Treatise on Nervous Exhaustion: Its Symptoms, Nature, Sequences, Treatment*, E. B. Treat, 1889.

16 It seems that the psychologist William James nicknamed neurasthenia 'americanitis'.

17 Freud, S., *Studies on Hysteria* (trans. J. Stratchey) (originally published 1895), Basic Books, 1957.

18 Klein, D. F., 'Delineation of two drug responsive anxiety syndromes', *Psychopharmacologia*, 5 (1964), 397–401.

19 American Psychiatric Association, *Diagnostic and Statistical Manual of Mental Disorders*, 4th edn, Text Revision (DSM-IV TR), American Psychiatric Press, 2000.

20 Adolphs, R., Tranel, D., Damasio, H., and Damasio, A., 'Impaired recognition of emotion in facial expressions following bilateral damage to the human amygdala', *Nature*, 372 (1995), 669–72.

21 I am enormously indebted to Simon Critchley for making me understand Heidegger through his writing. I have benefited greatly from his explanations of *Being and Time* in the *Guardian* in 2009. Any misinterpretation of Heidegger's thoughts is, of course, mine. Heidegger's original words are quoted from section 40 of chapter 6 of *Being and Time* (originally published 1927; republished by Harper & Row, 1962) and from Heidegger's beautiful essay 'What is metaphysics?'. Delivered as a lecture at Freiburg University in 1929, this describes anxiety's role in our lives, and also outlines how, according to Heidegger, the methods of science fail to understand our existence.

22 Heidegger, *Being and Time*, ch. 6 section 40.

23 'in der Angst ist einem unheimlich'.

24 Joseph LeDoux's work on the emotion of fear and anxiety has been pioneering. Research in his laboratory has unveiled much of what we have recently learnt about fear conditioning and the brain tissues involved. The redirection of the signal pathway of fear is described in Amorapanth, P., LeDoux, J. E., and Nader, K., 'Different lateral amygdala outputs mediate reactions and actions elicited by a fear-arousing stimulus', *Nature Reviews Neuroscience*, 3 (2000), 74–9.

25 Gozzi, A., Jain, A., Giovanelli, A., *et al.*, 'A neural switch for active and passive fear', *Neuron* 67 (2010), 656–66.

26 LeDoux, J., and Gorman, J. M., 'A call to action: Overcoming anxiety through active coping', *American Journal of Psychiatry*, 158 (2001), 1953–5. This article was written in the wake of September 11.

27 Knowledge of any kind can be harvested to inform resolute actions or generally improve our life, but sometimes we overlook its value. When interviewed by the online magazine *Slate* on how neuroscience has changed his life, Joseph LeDoux responded that one of the things he has learnt from his study of fear is that anxiety triggers anxiety and that breathing exercises such as those used in meditation are effective in dispelling anxiety, as they reduce the general arousal of the body. But he also admitted that he doesn't practise those exercises as often as he would like. See http://www.slate.com/articles/life/brains/2007/04/brain_lessons.html

28 Most probably such comparison would have not pleased Heidegger, I believe, since he dismissed science almost entirely. In 'What is metaphysics?' he wrote: '. . . no amount of scientific rigour attains to the seriousness of metaphysics. Philosophy can never be measured by the standard of the idea of science.'

29 The story is reported in 'After shock', *Guardian*, 17 June 2006.

30 Fakra, E., Hyde, L. W., Gorka, A., Fisher, P. M., Munoz, K. E., Kimak, M., Halder, I., Ferrell, R. E., Manuck, S. B., and Hariri, A. R., 'Effects of Htr1a C(-1019) G on amygdala reactivity and trait anxiety', *Archives of General Psychiatry*, 66 (2009), 33–40.

31 'From describing to nudging: Choice of transportation after a terrorist attack in London', a study of the impact of the July bombings on Londoners' travel behaviour: http://research.create.usc.edu/project_summaries/67

32 For a rich source on the plasticity of the brain, also in the cause of trauma, see the excellent Doidge, Norman, *The Brain That Changes Itself*, Penguin, 2007.

33 Beutel, M. E., Stark, R., Pan, H., Silbersweig, D., and Dietrich, S., 'Changes of brain activation pre-post short-term psychodynamic

inpatient psychotherapy: An fMRI study of panic disorder patients', *Psychiatry Research*, 184 (2010), 96–104.

34 'Medicine: To Nirvana with Miltown', *Time*, 7 July 1958.

35 For a comprehensive history of minor tranquillizers in the US, see Tone, Andrea, *The Age of Anxiety: A History of America's Turbulent Affair with Tranquilizers*, Basic Books, 2008.

36 For a rich review of the drug culture in the 1950s and a discussion of drug advertisements, especially in the US, see Metzl, Jonathan Michel, *Prozac on the Couch: Prescribing Gender in the Era of Wonder Drugs*, Duke University Press, 2003.

37 Smith, M., *Small Comfort: A History of the Minor Tranquilizers*, Praeger, 1985.

38 Lennard, H. L., Epstein, L. J., Bernstein, A., and Ranson, D. C., 'Hazards implicit in prescribing psychoactive drugs', *Science*, 169 (1970), 438–41.

39 Auden, *The Age of Anxiety*, p. 5.

40 I have also written about these issues in Frazzetto, G., 'Genetics of behavior and psychiatric disorders: From the laboratory to society and back', *Current Science*, 97 (2009), 1555–63. For a more extensive discussion of the relevance of Canguilhem's ideas in light of advances in the life sciences, see Rose, N., 'Life, reason and history: Reading Georges Canguilhem today', *Economy and Society*, 27 (1998), 154–70, and Canguilhem, G., *The Normal and the Pathological*, Zone Books, 1991.

41 A discussion of how contemporary society incites us to action, achievement and self-realization and has become less tolerant of mild anxious states, and of how individuals have become 'neurochemical selves' is presented in Rose, N., 'Neurochemical selves', *Society*, 41 (2003), 46–59.

42 A good argument in favour of this point can be found in Salecl, R., *On Anxiety*, Routledge, 2004.

43 Rilke, R. M., *Letters to a Young Poet*, W. W. Norton, 1993.

**Chapter 4**

1 These are the closing lines of Borges's poem 'Ausencia', meaning 'absence'. The translation from the Spanish is mine.

2 Kübler-Ross, E., and Kessler, D., *On Grief and Grieving*, Scribner, 2007.

3 The first use of the expression 'down in the mouth' is attributed to Bishop Joseph Hall in his *Resolutions and Decisions of Diverse Practical Cases of Conscience* (1649), cited in Rogers, James, *Dictionary of Clichés*, Wing Books, 1970.

4 Darwin, C., *The Expression of the Emotions in Man and Animals* (originally published 1872), in Wilson, E. O. (ed.), *From So Simple a Beginning: The Four Great Books of Charles Darwin*, Norton, 2006, p. 1362.

5 Ibid.

6 Ibid. p. 1348.

7 Provine, R. R., Krosnowski, K. A., and Brocato, N. W., 'Tearing: Breakthrough in human emotional signaling', *Evolutionary Psychology*, 7 (2009), 52–6; Provine, R. R., *Curious Behavior: Yawning, Laughing, Hiccupping and Beyond*, Belknap Press (Harvard University Press), 2012.

8 Provine, Krosnowski and Brocato, 'Tearing: Breakthrough in human emotional signaling'.

9 Hasson, O., 'Emotional Tears as Biological Signals', *Evolutionary Psychology*, 7 (2009), 363–70.

10 However, this does not seem to be a universal fact. The 'healing' quality of a bout of crying really depends on the circumstances, on the reason behind the tears and on who sheds them. A survey of about three thousand crying experiences showed that, while the majority of people felt better after crying, a few experienced no improvement in their mood and others felt actually worse; Bylsma, L. M., Vingerhoets, A. J. J. M., and Rottenberg, J., 'When is crying cathartic? An international study', *Journal of Social and Clinical Psychology*, 27 (2008), 1165–87.

11 Provine, R. R., 'Emotional tears and NGF: A biographical appreciation and research beginning', *Archives Italiennes de Biologie*, 149 (2011), 269–74.

12 Eisenberger, N. I., and Lieberman, M. D., 'Why it hurts to be left out. The neurocognitive overlap between physical and social pain', in Williams, K. D., Forgas, J. P., and von Hippel, W. (eds), *The Social Outcast: Ostracism, Social Exclusion, Rejection, and Bullying*, Cambridge University Press, 2005, pp. 109–27.

13 Panksepp, J., *Affective Neuroscience*, Oxford University Press, 1998.

14 Eisenberger, N. I., Lieberman, M. D., and Williams, K. D., 'Does rejection hurt? An fMRI study of social exclusion', *Science*, 302 (2003), 290–2.

15 O'Connor, M. F., Wellisch, D. K., Stanton, A. L., Eisenberger, N. I., Irwin, M. R., and Lieberman, M. D., 'Craving love? Enduring grief activates brain's reward center', *Neuroimage*, 42 (2008), 969–72.

16 Following the evolution of psychiatric thinking and advances in clinical and neurological research, depression assumed different forms and was

given various names, including 'involutional melancholia', 'depressive reaction', 'manic depressive illness' and 'depressive neurosis'. See Gruenberg, A. M., Goldstein, R. D., and Pincus, H. A., 'Classification of depression: Research and diagnostic criteria: DSM-IV and ICD-10', in Licinio, J., and Wong, M. L. (eds), *Biology of Depression: From Novel Insights to Therapeutic Strategies*, Wiley-VCH Verlag, 2005. For a comprehensive history of the concept of depression, see Jackson, S. W., *Melancholia and Depression: From Hippocratic Times to Modern Times*, Yale University Press, 1986.

17 Freud, S., *Mourning and Melancholia* (originally published 1917), 14th edn, Vintage, 1998.

18 American Psychiatric Association, *Diagnostic and Statistical Manual of Mental Disorders*, 4th edn, Text Revision (DSM-IV TR), American Psychiatric Press, 2000.

19 The American Psychiatric Association has published proposals and preliminary drafts for the fifth edition of the DSM on www.dsm5.org

20 Kendler, K. S., Myers, J. M. S., and Zisook, S., 'Does bereavement-related major depression differ from major depression associated with other stressful life events?', *American Journal of Psychiatry*, 165 (2008), 1449–55.

21 Prigerson, H. G., Horowitz, M. J., Jacobs, S. C., *et al.*, 'Prolonged Grief Disorder: Psychometric validation of criteria proposed for DSM-V and ICD-11', *PLOS Medicine*, 6 (2009), e1000121. For instance, the O'Connor *et al.* brain-imaging study that found similarities between pain and grief (see n. 15 above) was designed to identify the specific regions of activation in bereaved patients who had manifested severe symptoms of depression, and compare them with the regions of activation in a group manifesting milder symptoms. The authors argue that in patients with severe symptoms there is higher activity in the nucleus accumbens, one of the tissues in the brain's reward system, which I will talk about in the chapter on joy. They interpret this to be the consequence of an unresolved yearning in connection with wishful reveries about the deceased spouse that is not necessarily useful in coping with the loss or in paving the way to acceptance of the spouse's death.

22 Bromet, E., *et al.*, *BMC Medicine*, 9 (2011), 90. These figures report the number of people who have actually been diagnosed and are under observation or being treated, either through psychotherapy or with drugs or both.

23 Source: http://www.cdc.gov/nchs/data/nvsr/nvsr61/nvsr61_06.pdf.

24 See for instance: http://www.psychologytoday.com/blog/dsm5-in-distress/201008/good-grief-vs-major-depressive-disorder. Also:

Frances, Allen, *Saving Normal: An Insider Revolts against Out-of-control Psychiatric Diagnosis, DSM-5, Big Pharma, and the Medicalization of Everyday Life*, William Morrow, 2013.

25 For a moving appeal against the inclusion of grief in the DSM, see Arthur Kleinman's editorial in the *Lancet*, 379 (18 February 2012), 608–9.

26 I read this quote in an article by Zadie Smith on joy. Apparently Barnes heard this statement from a friend of his who wrote it in a letter of condolence; Smith, Z., 'Joy', *New York Review of Books*, 10 January 2013.

27 Wittgenstein, L., *Philosophical Investigations* (trans. G. E. M. Anscombe), Oxford University Press, 1953.

28 For a more extensive background to Wittgenstein and emotions, see Mascolo, M. F., 'Wittgenstein and the discursive analysis of emotion', *New Ideas in Psychology*, 27 (2009), 258–74.

29 Wittgenstein, *Philosophical Investigations*, #66. It is remarkable that Wittgenstein, at least in an early phase of his thinking, had a strong anti-naturalist attitude and completely turned his back on Darwin. For the Austrian thinker, the theory of evolution was not sufficient to explain the 'multiplicity' of species in the world. This is made clear in a conversation by letter with Maurice Drury: Drury, M. O'C., 'Conversations with Wittgenstein', in *Ludwig Wittgenstein: Personal Recollections*, ed. R. Rhees, Rowman and Littlefield, 1981. However, later, in *Philosophical Investigations*, Wittgenstein gave more importance to the body to account for the operations of the mind and he clearly accorded facial expressions great power in conveying emotion.

30 This quote is not from *Philosophical Investigations*, but is #570 in Wittgenstein, L., *Remarks on the Philosophy of Psychology*, Vol. II (trans. G. E. M. Anscombe), University of Chicago Press, 1980.

31 The drawings and quote I use here are from Wittgenstein, L., *Lectures on Aesthetics* (originally published 1966), University of California Press, 40th Anniversary Edition, 2007.

32 An emoticon (the word combines 'emotion' and 'icon') accompanies text to add an emotional tone to a message that could otherwise be misinterpreted.
Many attribute the first use of an emoticon to Professor Scott Fahlman, a computer scientist at Carnegie Mellon University, who on 19 September 1982 wrote :-) in an email to his colleagues at the university, followed by the comment: 'Read it sideways.' His aim was to help his college community to achieve the right tone in written communications, so as to avoid misunderstandings. However, the American commercial

artist Harvey Ball had designed the smiley ☺ in 1963. An even earlier use of print characters to mimic a facial expression was found in an 1862 transcript of a speech by President Abraham Lincoln. In the transcript, the characters ;) appear after the word 'laugh'. This may have been a typo, but if not, it is the oldest known documented use of an 'emoticon'. Emoticons add the visual elements that language inevitably omits. They now abound in virtual communications. When I send a text message with my iPhone and use an emoticon app – the most popular one is Emoji – to accompany or replace words, I am at a loss to know what all those faces mean. I have counted about sixty different images of faces, half of which seem to portray positive expressions and the other half negative ones. There are more emoticons than I can find words to describe them. Does this mean that we can emote in a more sophisticated manner than we used to, or that our language and its public use are richer?

Source on the history of emoticons: Lee, Jennifer, 'Is that an emoticon in 1862?', *New York Times*, 19 January 2009.

33 Hahn, T., 'Integrating neurobiological markers of depression: An fMRI based pattern classification approach', PhD thesis, Julius-Maximilians-Universität Würzburg, 2010.

34 For a short introduction on research-based diagnostic criteria, see a text by Thomas Insel, current director of the US National Institute of Mental Health: http://www.nimh.nih.gov/about/director/directors-biography.shtml

35 Pies, R., 'Why psychiatry needs to scrap the DSM system: An immodest proposal', http://psychcentral.com/blog/archives/2012/01/07/why-psychiatry-needs-to-scrap-the-dsm-system-an-immodest-proposal/

36 Pletscher, A., Shore, P. A., and Brodie, B. B., 'Serotonin release as a possible mechanism of reserpine action', *Science*, 122 (1955), 374–5. For a comprehensive history of antidepressants, see Healy, D., *The Anti-depressant Era*, Harvard University Press, 1997.

37 Schildkraut, J. J., 'The catecholamine hypothesis of affective disorders: A review of the supporting evidence', *American Journal of Psychiatry*, 122 (1965), 509–22.

38 The serotonin re-uptake mechanism was discovered by Sir Bernard Katz, Ulf von Euler and Julius Axelrod, who earned the Nobel Prize in Medicine for it in 1970: http://www.nobelprize.org/nobel_prizes/medicine/laureates/1970/

39 Data from IMS Health Inc, Biopharma Forecasts and Trends, as reported in Insel, T. R., 'Next-generation treatments for mental disorders', *Science Translational Medicine*, 4, Issue 155 (2012), 1–9.

40 Wittchen, H. U., Jacobi, F., Rehm. A., Gustavsson, A., Svensson, M., Jönsson, B., Olesen, J., Allgulander, C., Alonso, J., Faravelli, C., Fratiglioni, L., Jennum, P., Lieb, R., Marcker, A., *et al.*, 'The size and burden of mental disorders and other disorders of the brain in Europe 2010', *European Neuropsychopharmacology*, 21 (2011), 655–79.

41 Lacasse, J. R., and Leo, J., 'Serotonin and depression: A disconnect between the advertisements and the scientific literature', *PLOS Medicine*, 2, Issue 12 (2005), e392. For a sharp analysis of how ambiguity and uncertainty around the efficacy of antidepressants can be employed to fortify their perceived authority, see McGoey, L., 'On the will to ignorance in bureaucracy', *Economy and Society*, 36 (2007), 212–35, and McGoey, L., 'Profitable failure: Antidepressant drugs and the triumph of flawed experiments', *History of the Human Sciences*, 23 (2010), 58–78.

42 Neville, S., 'GlaxoSmithKline fined $3bn after bribing doctors to increase drug sales', *Guardian*, 3 July 2012.

43 Kirsch, I., Deacon, B. F., Huedo-Medina, T. B., *et al.*, 'Initial severity and antidepressant benefits: A meta-analysis of data submitted to the Food and Drug Administration', *PLOS Medicine*, 5, Issue 2 (2008), e45.

44 Miller, G., 'Is Pharma running out of brainy ideas?', *Science*, 329 (2010), 502–4.

45 For a review of current directions in the discovery of new treatments for mental disorders, see Insel, 'Next-generation treatments for mental disorders'.

46 For my summary of the theory of the humours, I have consulted the excellent history of the theory by Noga Arikha: *Passions and Tempers: A History of the Humours*, Ecco, 2007.

47 Lloyd, G. E. R. (ed.), *Hippocratic Writings* (trans. J. Chadwyck and W. N. Mann), Penguin, 1978.

48 Ibid.

49 Arikha, *Passions and Tempers*. One such treatise is *The Treatise of Fevers* by the Arabic doctor Ishaq ben Sulayman al-Israeli. In Chapter XI al-Israeli talks about a fever caused by sorrow, which was governed by the emotions and was characterized by agitation and lack of appetite: Sluzki, C., 'On Sorrow: Medical advice from Ishaq ben Sulayman al-Israeli, 1000 years ago', *American Journal of Psychiatry*, 167 (2010), 5. I also thank Dr Sluzki for kindly providing me with an earlier, longer version of his paper on the treatise.

50 Arikha, *Passions and Tempers*; Sluzki, 'On Sorrow: Medical advice from Ishaq ben Sulayman al-Israeli, 1000 years ago'.

51 'Living with grief', editorial, *Lancet*, 379 (2011), 589.

52 I read Robert Pinsky's poem 'Grief' – which also inspired the subtitle of this chapter – in the *New York Review of Books*, 7 June 2012.

**Chapter 5**

1 Darwin, C., *The Expression of the Emotions in Man and Animals* (originally published 1872), in Wilson, E. O. (ed.), From *So Simple a Beginning: The Four Great Books of Charles Darwin*, Norton, 2006, p. 1476.

2 Ibid. p. 1434.

3 Ibid. p. 1402.

4 Ibid. p. 1420.

5 For a good general review of the function of fiction in simulating social experience, see Mar, R. A., and Oatley, K., 'The function of fiction is the abstraction and simulation of social experience', *Perspectives on Psychological Science*, 3 (2008), 173–92.

6 Titchener, E. B., *Lectures on the Experimental Psychology of Thought Processes*, Macmillan, 1909; Vischer, R., *Über das optische Formgefühl: Ein Beiträg zur Ästhetik*, Credner, 1873; Lipps, T., *Grundlegung der Aesthetik*, Engelmann, 1903.

7 For a review of the role of empathy in aesthetic experience, see: Freedberg, D., and Gallese, V., 'Motion, emotion and empathy in esthetic experience', *Trends in Cognitive Sciences*, 11 (2007), 197–203.

8 For a definition and in-depth study of empathy, as a capacity to understand what another person is thinking or feeling, see Baron-Cohen, S., *Zero Degree of Empathy*, Penguin (Allen Lane), 2011.

9 Certainly theatre is not the only vehicle for empathic communication with a fictional world. Reading a book or watching a film at the cinema may both evoke powerful emotions. However, in theatre we have a *live* demonstration of emotions, like the demonstration of an experiment in the laboratory. Reading a book – except in the case of public readings – is mostly a private affair, whereas watching a performance is a collective action. When we read a novel, a short story or a play, we create in our mind coherent representations of the characters, places and imagery presented in the work, from start to end. Through the author's words, the characters take shape and possess distinct qualities and features. Each assumes a physical appearance and, in our mind's eye, takes on a life of his or her own. Even after we have finished reading the work, the characters maintain the appearances acquired while we read about them – until, of course, cinema adapts the book for the screen and gives them the faces of famous movie stars.

In cinema, suggestive music, close-ups and particularly camera angles can immensely enhance our emotional responses. Yet the story is displayed on the fixed and physically constrained surface of the screen. So we are offered a flat, two-dimensional representation of a story. The acting we see is over by the time it is edited and watched in cinemas. In theatre, by contrast, stories are told and acted by flesh-and-blood actors in a time and space shared with an audience who may influence the delivery of the performance. In this way, theatre is a unique space in which to connect and share an experience.

10 Ramon y Cajal, S., *Advice for a Young Investigator*, MIT Press, 1999.

11 For a review of mirror neurons, see Rizzolatti, G., and Craighero, L., 'The mirror-neuron system', *Annual Review of Neuroscience*, 27 (2004), 169–92, and Gallese, V., 'The roots of empathy: The shared manifold hypothesis and the neural basis of intersubjectivity', *Psychopathology*, 36 (2003), 171–80.

12 Gallese, V., Fadiga, L., and Rizzolatti, G., 'Action recognition in the premotor cortex', *Brain*, 119 (1996), 593–609; Rizzolatti, G., Fadiga, L., Gallese, V., and Fogassi, L., 'Premotor cortex and the recognition of motor actions', *Cognitive Brain Research*, 3 (1996), 131–41.

13 The fact that the mirroring type of firing that takes place when observing an action does not result in any actual motor action may be due to an incomplete or insufficient pattern of firing, or to some subsequent inhibition mechanism that prevents the execution of the action.

14 Iacoboni, M., *et al.*, 'Cortical mechanisms of human imitation', *Science*, 286 (2003), 2526–8.

15 Chakrabarti, B., Bullmore, E., and Baron-Cohen, S., 'Empathising with basic emotions: Common and discrete neural substrates', *Social Neuroscience*, 1 (2006), 364–84.

16 Carr, L., Iacoboni, M., Dubeau, M. C., Mazziotta, J. C., and Lenzi, G. L., 'Neural mechanisms of empathy in humans: A relay from neural systems for imitation to limbic areas', *Proceedings of the National Academy of Sciences*, 100 (2003), 5497–502.

17 Wicker, B., Keysers, C., Plailly, J., Royet, J. P., Gallese, V., and Rizzolatti, G., 'Both of us disgusted in my insula: The common neural basis of seeing and feeling disgust', *Neuron*, 40 (2003), 655–64.

18 Keysers, C., Wicker, B., Gazzola, V., Anton, J. L., Fogassi, L., and Gallese, V., 'A touching sight: SII/PV activation during the observation and experience of touch', *Neuron*, 42 (2004), 335–46.

19 Haker, H., Kawohl, W., Herwig, U., and Rössler, W., 'Mirror neuron activity during contagious yawning – an fMRI study', *Brain Imaging and Behavior* (7 July 2012: electronic abstract ahead of publication).

20 Brook, P., *The Empty Space* (originally published 1968), Penguin, 1990.

21 Based on an interview with Ben Crystal and on his book *Shakespeare on Toast*, Icon Books, 2009.

22 Shaw, G. B., *Our Theatres in the Nineties*, Vol. I of *Collected Works*, Constable and Company, 1932, p. 154.

23 Ibid.

24 Stanislavsky, C., *Creating a Role*, Methuen, 1961, p. 106.

25 Stanislavsky, C., *An Actor Prepares*, Theatre Art Books, 1936 (1983 printing).

26 Ibid. p. 121.

27 Stanislavsky most probably adopted the concept of emotional memory from the French scientist Théodule Ribot (1839–1916), who first used the term 'affective' memory. Ribot is mentioned in Stanislavsky's *An Actor Prepares* to illustrate how we hardly forget the emotions we feel (p. 156).

28 Stanislavsky, *An Actor Prepares*, p. 159.

29 Ibid. p. 164.

30 Ibid. p. 158.

31 Ibid. p. 163.

32 Sawoski, P., *The Stanislavsky System: Growth and Methodology*, Teaching Material, Santa Monica College, Spring 2010.

33 Stanislavsky, *An Actor Prepares*, p. 133.

34 Ibid. p. 122.

35 Ibid. p. 123.

36 Ibid. p. 47.

37 Interview by Lynn Hirschberg, *Daily Telegraph*, 8 December 2007.

38 Diderot, D., *The Paradox of Acting* (trans. Walter H. Pollock), Chatto and Windus, 1883.

39 Ibid. p. 8.

40 Ibid. p. 9.

41 Ibid. p. 39.

42 There is often confusion about what is meant by 'method acting' and about the attribution to Stanislavsky of this approach to dramatic performance. It is commonly accepted that the American method of acting is an echo of the Russian director's teachings on emotional memory. The reason why Stanislavsky's later focus on physical actions did not enter the American theatrical tradition guided by personalities like Lee Strasberg may be a chronological accident. Stanislavsky described the evolution of his method in his books published in 1936 and 1949, by which time the 'method' based on portraying a character 'inside-out' had already become routine; Gray, P., 'From Russia to America: A critical chronology' in Munk, E. (ed.), *Stanislavsky and America*, Hill and Wang, 1966. See also Sawoski, *The Stanislavsky System: Growth and Methodology.*

43 Stanislavsky, *An Actor Prepares*, p. 123.

44 Diderot, *The Paradox of Acting*, p. 18.

45 A recent study explored how spectators reacted to an identical scene played in two different ways: one played by actors who 'incarnate' the role and one by actors who are more detached. In the end, the spectators judged the actors with the detached approach to be closer to the character of their role and showing more powerful emotions: Goldstein, T., 'Responses to and judgments of acting on film', in Kaufman, J. C., and Simonton, D. K. (eds), *The Social Science of Cinema*, Oxford University Press, 2012.

46 Interview by Charles McGrath, *New York Times*, 31 October 2012.

47 Abraham, A., von Cramon, D. Y., and Schubotz, R. I., 'Meeting George Bush versus Cinderella: The neural response when telling apart what is real from what is fictional in the context of our reality', *Journal of Cognitive Neuroscience*, 20 (2008), 965–76. I am grateful to Dr Anna Abraham for explaining her work to me during an interview.

48 Abraham, A., and von Cramon, D. Y., 'Reality-relevance? Insights from spontaneous modulations of the brain's default network when telling apart reality from fiction', *PLOS One*, 4 (2009), e4741.

49 Coleridge, Samuel Taylor, *Biographia Literaria*, Chapter XIV, 1817.

50 Directors come up with countless ways to either induce or suspend disbelief. I once watched an unforgettable show by the Théâtre du Soleil, led by the French director Ariane Mnouchkine. To reach my seat I had to walk through a corridor where on one side, separated by a see-through curtain, actors were putting on their make-up. Through this unusual tactic – which I interpreted as a way to bring spectators close to, and familiarize them with, the actors – the cast were whispering to us: we are going to fake a story. Of course, I knew that whatever I was about to watch was fiction. I spent a few minutes looking at the cast. I recognized some of them when they showed up in character on stage. Yet, as the show unfolded, this

awareness did not interfere with the power of the story and the actors' skill in making me forget that the characters were actors in disguise.

51 Brecht, B. (ed. and trans. John Willett), *Brecht on Theatre*, 2nd edn, Methuen, 2001; Freshwater, H., *Theatre and the Audience*, Palgrave Macmillan, 2009.

52 Interview by Robert Ayers: http://www.askyfilledwithshootingstars.com/wordpress/?p=1197

53 For a review of the relationship between direct eye contact and the social brain network, see Senju, A., and Johnson, M. H., 'The eye contact effect: Mechanisms and development', *Trends in Cognitive Sciences*, 13 (2009), 127–34. One area in particular that is involved when we follow the direction of someone else's eyes is the posterior superior temporal sulcus (pSTS). A person in whom this area is damaged or missing finds it hard to assess accurately where someone else is gazing or interpret what they are feeling about the object they are gazing at. See Campbell, R., Heywood, C., Cowey, A., Regard, M., and Landis, T., 'Sensitivity to eye gaze in prosopagnosic patients and monkeys with superior temporal sulcus ablation', *Neuropsychologia*, 28 (1990), 1123–42. As I explain later, the pSTS is also activated when we watch a play. See Metz-Lutz, M. N., Bressan, Y., Heider, N., and Otzenberger, H., 'What physiological and cerebral traces tell us about adhesion to fiction during theater-watching', *Frontiers in Human Neuroscience*, 4 (2010), Article 59, 1–10.

54 Kampe, K., Frith, C. D., Dolan, R. J., and Frith, U., 'Reward value of attractiveness and gaze', *Nature*, 413 (2001), 589.

55 The original title is *Onysos le furieux*, a play written by the French playwright Laurent Gaude.

56 Metz-Lutz, Bressan, Heider, and Otzenberger, 'What physiological and cerebral traces tell us about adhesion to fiction during theater-watching'. I have previously commented on this work in Frazzetto, G., 'Powerful Acts', *Nature*, 482 (2012), 466–7. I am grateful to Dr Yannick Bressan for discussing his work with me in an email exchange.

57 Jabbi, M., and Keysers, C., 'Inferior frontal gyrus activity triggers anterior insula response to emotional facial expressions', *Emotion*, 8 (2008), 775–80.

58 Campbell, Heywood, Cowey, Regard, and Landis, 'Sensitivity to eye gaze in prosopagnosic patients and monkeys with superior temporal sulcus ablation'.

59 Rapp, A., Leube, D. T., Erb, M., Grodd, W., and Kircher, T. T., 'Neural correlates of metaphor processing', *Cognitive Brain Research*, 20 (2004), 395–402.

60 Zysset, S., Huber, O., Ferstl, E., and von Cramon, D. Y., 'The anterior frontomedian cortex and evaluative judgment: An fMRI study', *Neuroimage* 15 (2002), 983–91.

61 Brook, *The Empty Space*.

**Chapter 6**

1 Freud, S., *Civilization and Its Discontents* (originally published 1930), Penguin, 2002.

2 Ekman, P., *Emotions Revealed*, Henry Holt and Company, 2003, p. 205; Darwin, C., *The Expression of the Emotions in Man and Animals* (originally published 1872), in Wilson, E. O. (ed.), *From So Simple a Beginning: The Four Great Books of Charles Darwin*, Norton, 2006.

3 Provine, R., 'Laughter', *American Scientist*, 84 (1996), 38–45; Provine, R. R., *Curious Behavior: Yawning, Laughing, Hiccupping and Beyond*, Belknap Press (Harvard University Press), 2012.

4 Burgdorf, J., Panksepp, J., Brudzynski, S. M., Kroes, R., and Moskal, J. R., 'Breeding for 50-kHz positive affective vocalization in rats', *Behavior Genetics*, 35 (2005), 67–72.

5 Sauter, D. A., Eisner, F., Ekman, P., and Scott, S. K., 'Cross-cultural recognition of basic emotions through emotional vocalizations', *Proceedings of the National Academy of Sciences*, 107 (2010), 2408–12.

6 Warren, J. E., Sauter, D. A., Eisner, F., *et al.*, 'Positive emotions preferentially engage an auditory-motor "mirror" system', *Journal of Neuroscience*, 26 (2006), 13,067–75.

7 Provine, R., and Fischer, K. R., 'Laughing, smiling and talking: Relation to sleeping and social context in humans', *Ethology*, 83 (1989), 295–305.

8 Hammer, M., 'An identified neuron mediates the unconditioned stimulus in associative olfactory learning in honeybees', *Nature*, 366 (1993), 59–63.

9 Olds, J., and Milner, P., 'Positive reinforcement produced by electrical stimulation of septal area and other regions of rat brain', *Journal of Comparative and Physiological Psychology*, 47 (1954), 419–28.

10 Schultz, W., Apicella, P., and Ljungberg, T., 'Responses of monkey dopamine neurons to reward and conditioned stimuli during successive steps of learning a delayed response task', *Journal of Neuroscience*, 13 (1993), 900–13; Schultz, W., Apicella, P., Scarnati, E., and Ljungberg, T., 'Neuronal activity in monkey ventral striatum related to the expectation of reward', *Journal of Neuroscience*, 12 (1992), 4595–610. For a review of predictive rewards, see Schultz, W., 'Multiple reward signals in the brain', *Nature Reviews Neuroscience*, 1 (2000), 199–207.

11 Fiorino, F., Coury, A., and Phillips, A. G., 'Dynamic changes in nucleus accumbens dopamine efflux during the Coolidge effect in male rats', *Journal of Neuroscience*, 17 (1997), 4849–55.

12 Hammer, M., and Menzel, R., 'Learning and memory in the honeybee', *Journal of Neuroscience*, 15 (1995), 1617–30.

13 The article exploring predictive learning in bees via computer simulation is: Montague, P. R., Dayan, P., Person, C., and Sejnowski, T. J., 'Bee foraging in uncertain environments using predictive Hebbian learning', *Nature*, 377 (1995), 725–8.

14 For a detailed review of how a loop of information is established between pleasure centres and the prefrontal cortex, and how abstract thoughts can be built from simpler ones, see Miller, E. K., and Buschman, T. J., 'Rules through recursion: How interactions between the frontal cortex and basal ganglia may build abstract, complex rules from concrete, simple ones', in Bunge, S., and Wallis, J. (eds), *Neuroscience of Rule-Guided Behavior*, Oxford University Press, 2007, pp. 419–40.

15 For a very informative review of the effect of positive feelings on working memory and for a description of the neurobiology of problem solving, see Ashby, F. G., Valentin, V. V., and Turken, U., 'The effects of positive affect and arousal on working memory and executive attention', in Moore, S., and Oaksford, M. (eds), *Emotional Cognition: From Brain to Behavior*, John Benjamin Publishing, 2002, pp. 245–87.

16 Isen, A., Daubman, K. A., and Nowicki, G. P., 'Positive affect facilitates creative problem solving', *Journal of Personality and Social Psychology*, 52 (1987), 1122–31.

17 Isen, A. M., Johnson, M. M. S., Mertz, E., and Robinson, G. F., 'The influence of positive affect on the unusualness of word associations', *Journal of Personality and Social Psychology*, 48 (1985), 1413–26.

18 Ginsberg, A., *Howl, Kaddish and Other Poems*, Penguin Classics, 2009.

19 Sahakian, B., and Morein-Zamir, S., 'Professor's little helper', *Nature*, 450 (2007), 1157–9.

20 Maher, B., 'Poll results: Look who's doping', *Nature*, 452 (2008), 674–5.

21 Sample, Ian, 'Female orgasm captured in series of brain scans', *Guardian*, 14 November 2011.

22 Holstege, G., Georgiadis, J. R., Paans, A. M. J., *et al.*, 'Brain activation during human male ejaculation', *Journal of Neuroscience*, 23 (2003), 9185–93.

23 Komisaruk, B. R., and Whipple, B., 'Functional MRI of the brain during orgasm in women', *Annual Review of Sex Research*, 16 (2005),

62–86; Komisaruk, B. R., Whipple, B., Crawford, A., *et al.*, 'Brain activation during vaginocervical self-stimulation and orgasm in women with complete spinal cord injury: fMRI evidence of mediation by the Vagus nerves', *Brain Research*, 1024 (2004), 77–88.

24 Darwin, C., *The Descent of Man, and Selection in Relation to Sex* (originally published 1871), in Wilson, E. O. (ed.), *From So Simple a Beginning: The Four Great Books of Charles Darwin*, Norton, 2006. In footnote n. 944.

25 Goldstein, R., 'Thrills in response to music and other stimuli', *Physiological Psychology*, 8 (1980), 126–9.

26 Sloboda, J. A., 'Music structure and emotional response: Some empirical findings', *Psychology of Music*, 19 (1991), 110–20.

27 Blood, A. J., and Zatorre, R. J., 'Intensely pleasurable responses to music correlate with activity in brain regions implicated in reward and emotion', *Proceedings of the National Academy of Sciences*, 98 (2001), 11,818–23.

28 Goldstein, 'Thrills in response to music and other stimuli'.

29 For an excellent review of the distinction between wanting and liking in food reward, see Berridge, K. C., '"Liking" and "wanting" food rewards: Brain substrates and roles in eating disorders', *Physiology and Behavior*, 97 (2009), 537–50.

30 Woolley, J. D., and Fields, H. L., 'Nucleus accumbens opioids regulate flavor-based preferences in food consumption', *Neuroscience*, 17 (2006), 309–17.

31 Gainotti, G., 'Emotional behavior and hemispheric side of the lesion', *Cortex*, 8 (1972), 41–55; Sackeim, H. A., Greenberg, M. S., Weiman, A. L., *et al.*, 'Hemispheric asymmetry in the expression of positive and negative emotions', *Archives of Neurology*, 39 (1982), 210–18.

32 See Davidson, R. J. (and Begley, S.), *The Emotional Life of Your Brain*, Hudson Street Press (Penguin), 2012.

33 The original article describing the experiment with negative emotions is Schwartz, G. E., Davidson, R. J., and Maer, F., 'Right hemisphere lateralization for emotion in the human brain: Interactions with cognition', *Science*, 190 (1975), 286–8.

34 Davidson, R. J., and Fox, N. A. 'Asymmetrical brain activity discriminates between positive versus negative affective stimuli in human infants', *Science*, 218 (1982), 1235–7.

35 Davidson, R. J., Ekman, P., Saron, C. D., Senulis, J. A., and Friesen, W. V., 'Approach-Withdrawal and cerebral asymmetry: Emotional

expression and brain physiology I', *Journal of Personality and Social Psychology*, 58 (1990), 330–41.

36 Ekman, P., Davidson, R. J., and Friesen, W. V., 'The Duchenne smile: Emotional expression and brain physiology II', *Journal of Personality and Social Psychology*, 58 (1990), 342–53.

37 Davidson, R. J., and Fox, N. A., 'Frontal brain asymmetry predicts infants' response to maternal separation', *Journal of Abnormal Psychology*, 98 (1989), 127–31.

38 Schaffer, C. E., Davidson, R. J., and Saron, C., 'Frontal and parietal electroencephalogram asymmetry in depressed and nondepressed subjects', *Biological Psychiatry*, 18 (1987), 753–62.

39 His observation of the large individual variation in hemispheric activity in response to the same stimulus made Davidson come up with the idea that people have *emotional styles*. For a complete account of the evolution of this theory and its meaning, see Davidson (and Begley), *The Emotional Life of Your Brain*.

40 Foot, P., 'A new definition' (Recipes for Happiness), *British Medical Journal*, 321 (2000), 1576.

41 For an extensive, though rather academic, explanation of Mandeville's thinking, and especially his contribution to moral psychology and the science of human nature, see Hundert, E. J., *The Enlightenment's Fable: Bernard Mandeville and the Discovery of Society*, Cambridge University Press, 1994.

42 Indeed, it can be said that the Enlightenment was the stage for the first sexual revolution. A wonderful book documenting this is Dabhoiwala, F., *The Origins of Sex: A History of the First Sexual Revolution*, Penguin, 2013; for a short summary of the evolution of the concept of pleasure and hedonism in the Enlightenment, see Porter, R., 'Happy hedonists' (Recipes for Happiness), *British Medical Journal*, 321 (2000), 1572–5.

43 Harker, L. A., and Keltner, D., 'Expressions of positive emotion in women's college yearbook pictures and their relationship to personality and life outcomes across adulthood', *Journal of Personality and Social Psychology*, 80 (2001), 112–24.

44 Abel, E. L., and Kruger, M. L., 'Smile intensity in photographs predicts longevity', *Psychological Science*, 21 (2010), 542–4.

45 Keltner, D., and Bonanno, G. A., 'A study of laughter and dissociation: Distinct correlates of laughter and smiling during bereavement', *Journal of Personality and Social Psychology*, 4 (1997), 687–702.

46 Dunbar, R. I. M., Baron, R, Frangou, A., *et al.*, 'Social laughter is correlated with an elevated pain threshold', *Proceedings of the Royal Society B: Biological Sciences*, 279 (2011), 1161–7.

47 Cohen, S., Alper, C. M., Doyle, W. J., *et al.*, 'Positive emotional style predicts resistance to illness after experimental exposure to rhinovirus or influenza A virus', *Psychosomatic Medicine*, 68 (2006), 809–15.

48 Layard, R., *Happiness: Lessons from a New Science*, Penguin, 2005.

49 For a comprehensive review of wealth inequalities and happiness, see ibid., chapter 4. See also Diener, E., and Biswar-Diener, R., 'Will money increase subjective well-being?', *Social Indicators Research*, 57 (2002), 119–69.

50 Dunn, E. W., Aknin, L. B., and Norton, M. I., 'Spending money on others promotes happiness', *Science*, 319 (2008), 1687–8.

51 An important factor contributing to well-being is the ability to feel compassion. Though I do not report it here, scientists have begun to unravel some of the benefits of compassion for one's well-being as well as the transformations it brings about in the brain. One strategy for achieving this ability is meditation. Richard Davidson has extensively studied the signature, as it were, that meditation leaves on the brain and has shown that not only experienced meditators such as skilled Buddhist monks, but also ordinary people who approach meditation as beginners, are able to translate their practice into greater compassion towards family, friends and strangers, as well as sustain positive emotions and well-being. For the latest research on this, see Weng, H. Y., Fox, A. S., Shackman, A. J., Stodola, D. E., *et al.*, 'Compassion training alters altruism and neural responses to suffering', *Psychological Science*, in press 2013; another good source on meditation and the brain is Davidson (and Begley), *The Emotional Life of Your Brain*, chapter 10.

52 Holt-Lunstad, J., Smith, T. B., and Layton, J. B., 'Social relationships and mortality risk: A meta-analytic review', *PLOS Medicine*, 7, Issue 7 (2010), e1000316.

53 For a background review of the theory of the vagus nerve, see Porges, S. W., 'The polyvagal perspective', *Biological Psychology*, 74 (2007), 116–43.

54 Kok, B. E., and Fredrickson, B. L., 'Upward spirals of the heart: Autonomic flexibility, as indexed by vagal tone, reciprocally and prospectively predicts positive emotions and social connectedness', *Biological Psychology*, 85 (2010), 432–6.

55 Kok, B. E., Coffey, E. A., Cohn, M. A., *et al.*, 'How positive emotions build physical health: Perceived positive social connections account for

the upward spiral between positive emotions and vagal tone', *Psychological Science*, in press.

56 Baldwin, J., Letter: 'From a region in my mind', *New Yorker*, 17 November 1962.

57 The psychologist Mihaly Csikszentmihalyi has ascribed the term 'flow' to the experience of losing yourself in what you do. Other psychologists, such as Martin Seligman, call this strategy 'engagement'. See Peterson, C., Nansook, P., and Seligman, M. E. P., 'Orientations to happiness and life satisfaction: The full life versus the empty life', *Journal of Happiness Studies*, 6 (2005), 25–41.

**Chapter 7**

1 Because of love's complexity and various stages, not everyone agrees that it can be regarded as an emotion per se. However, love remains what most people intuitively refer to as one of the most common emotions.

2 Tomlinson, Simon, *Daily Mail*, 11 December 2012.

3 Citations of Plato's works are from Reeve, C. D. C., *Plato on Love*, Hackett Publishing Company, 2006.

4 English translation of Jacopo's sonnet in Stewart, D. E., *The Arrow of Love*, Associated University Presses, 2010.

5 Bartels, A., and Zeki, S., 'The neural basis of romantic love', *Neuroreport*, 11 (2000), 3829–34; Bartels, A., and Zeki, S., 'The neural correlates of maternal and romantic love', *Neuroimage*, 21 (2004), 1155–66.

6 Aron, A., Fisher, H., Mashek, D. J., Strong, G., Li, H., and Brown, L. L., 'Reward, motivation and emotion systems associated with early-stage intense romantic love', *Journal of Neurophysiology*, 94 (2005), 327–37.

7 The questionnaire used is the Passionate Love Scale; Hatfield, E., and Sprecher, S., 'Measuring passionate love in an intimate relation', *Journal of Adolescence*, 9 (1986), 383–410.

8 It is noteworthy that, straight or gay, love is one and universal. The brain makes no distinction between heterosexual and homosexual love relationships. Indeed, the imaging studies that have identified the reward/motivational system of the brain as part of the neural components of romantic love have been replicated and have yielded the same results among homosexuals, both men and women; Zeki, S., and Romaya, J. P., 'The brain reaction to viewing faces of opposite- and same-sex romantic partners', *PLOS One*, 5, Issue 12 (2010), e15802.

9 Stendhal, *On Love* (originally published 1822), Penguin, 1975, 2004.

10 Zeki, S., 'The neurobiology of love', *FEBS Letters*, 581 (2007), 2575–9.

11 Barthes, R., *A Lover's Discourse*, Hill and Wang, 1978, p. 133.

12 Zeki, 'The neurobiology of love'.

13 Capgras, J., 'L'illusion des "sosies" dans un délire systématisé chronique', *Bulletin de la Société Clinique de Médecine Mentale*, 11 (1923), 6–16; Hirstein, W., and Ramachandran, V. S., 'Capgras syndrome: A novel probe for understanding the neural representation of the identity and familiarity of persons', *Proceedings of the Royal Society B: Biological Sciences*, 264 (1997), 437–44; Debruille, J. B., and Stip, E., 'Capgras syndrome: Evolution of the hypothesis', *Canadian Journal of Psychiatry*, 41 (1996), 181–7.

14 The translation from the original clinical report in French is mine. See Capgras, J., 'L'illusion des "sosies" dans un délire systématisé chronique'.

15 See for instance Ramachandran, V. S., *The Tell-Tale Brain*, William Heinemann, Random House, 2011. Oliver Sacks has also been interested in the Capgras syndrome, studying it alongside prosopagnosia, the inability to recognize faces: Sacks, O., *The Mind's Eye*, Picador, 2010. The Capgras syndrome is also the subject of novels such as *The Echo Maker* by Richard Powers and *Atmospheric Disturbances* by Rivka Galchen.

16 With director Sommer Ulrickson, I staged a play called *Never Mind*, which was based on a text I wrote with the actors that was inspired by the Capgras syndrome. It premiered in Berlin on 25 January 2012 at the Sophiensäle: http://www.sophiensaele.com/produktionen.php?IDstueck=901

17 An excellent source covering all aspects of love, and not only the early romantic passion, is Appignanesi, L., *All about Love. Anatomy of an Unruly Emotion*, Virago Press, 2011.

18 For a review of oxytocin, vasopressin and attachment, see Insel, T. R., 'The challenge of translation in social neuroscience: A review of oxytocin, vasopressin, and affiliative behavior', *Neuron*, 65 (2010), 768–79; for the original paper on the distribution of the oxytocin receptors, see Insel, T., 'Oxytocin – a neuropeptide for affiliation: Evidence from behavioural, receptor autoradiographic, and comparative studies', *Psychoneuroendocrinology*, 17 (1992), 3–35.

19 Walum, H., Westberg, L., Henningsson, Jenae M., *et al.*, 'Genetic variation in the vasopressin receptor 1a gene (AVPR1A) associates with pair-bonding behavior in humans', *Proceedings of the National Academy of Sciences*, 105 (2008), 14,153–6.

20 Larkin, P., *The Complete Poems*, Faber and Faber, 2012. Reprinted with permission.

21 Bowlby, J., *Attachment and Loss*, Vol. II, Basic Books, 1973, p. 235.

22 For an article on romantic love and attachment, see Hazan, C., and Shaver, P., 'Romantic love conceptualized as an attachment process', *Journal of Personality and Social Psychology*, 52 (1987), 511–24.

23 Carola, V., Frazzetto, G., and Gross, C., 'Identifying interactions between genes and early environment in the mouse', *Genes, Brain and Behavior*, 5 (2006), 189–99. I am grateful to Valeria Carola for having taught me how to observe maternal behaviour in mice. She and I alternated the long hours of shifts in front of the mouse cages.

24 These data refer to access monitored in April 2011: Subscription Site Insider (2011) Dating and matchmaking site benchmark report. Newport, RI: Anne Holland Ventures; for a detailed, comprehensive review of the benefits and limitations of online dating see Finkel, E. J., Eastwick, P. W., Karney, B. R., Reis, H. T., and Sprecher, S., 'Online dating: A critical analysis from the perspective of psychological science', *Psychological Science in the Public Interest*, 13 (2012), 3–66.

25 Toma, C. L., Hancock, J. T., and Ellison, N. B., 'Separating fact from fiction: An examination of deceptive self-representation on online dating profiles', *Personality and Social Psychology Bulletin*, 34 (2008), 1023–36.

26 For an account of how love and relationships have become commodified, and the rise of 'emotional capitalism', see Illouz, Eva, *Cold Intimacies*, Polity, 2007, especially the chapter 'Romantic Webs'.

27 Fisher, H., *Why Him, Why Her*, One World Publications, 2009.

28 The most widely used and accepted inventory of personality identifies the 'Big Five' traits: Openness to new experiences, Conscientiousness, Extroversion, Agreeableness and Neuroticism, forming the acronym OCEAN; McRae, R. R., and Costa, P. T., 'Validation of the five-factor model across instruments and observers', *Journal of Personality and Social Psychology*, 52 (1987), 81–90.

## Epilogue

1 The rabbit–duck illusion is a multi-stable figure attributed to the American psychologist Joseph Jastrow and later used by Wittgenstein in Philosophical Investigations.

# Bibliography

## Books

Abad, H., Recipes for Sad Women (trans. Anne McLean), Pushkin Press, 2012.

American Psychiatric Association, *Diagnostic and Statistical Manual of Mental Disorders*, 4th edn, Text Revision (DSM-IV TR), American Psychiatric Press, 2000.

Appignanesi, L., *All about Love. Anatomy of an Unruly Emotion*, Virago Press, 2011.

Arikha, Noga, *Passions and Tempers: A History of the Humours*, Ecco, 2007.

Ashby, F. G., Valentin, V. V., and Turken, U., 'The effects of positive affect and arousal on working memory and executive attention', in Moore, S., and Oaksford, M. (eds), *Emotional Cognition: From Brain to Behavior*, John Benjamin Publishing, 2002.

Atik, A., *How It Was: A Memoir of Samuel Beckett*, Faber and Faber, 2001.

Auden, W. H., *The Age of Anxiety* (originally published 1947), Princeton University Press, 2011.

Baron-Cohen, S., *Zero Degree of Empathy*, Penguin (Allen Lane), 2011.

Barthes, R., *A Lover's Discourse*, Hill and Wang, 1978.

Beard, G. M. (with Rockwell, A. D.), 'Nervous exhaustion (neurasthenia)', Chapter I in *A Practical Treatise on Nervous Exhaustion: Its Symptoms, Nature, Sequences, Treatment*, E. B. Treat, 1889.

Bowlby, J., *Attachment and Loss*, Vol. II, Basic Books, 1973.

Brecht, B. (ed. and trans. John Willett), *Brecht on Theatre*, 2nd edn, Methuen, 2001.

Brook, P., *The Empty Space* (originally published 1968), Penguin, 1990.

Bunge, S., and Wallis, J. (eds), *Neuroscience of Rule-Guided Behavior*, Oxford University Press, 2007.

Canguilhem, G., *The Normal and the Pathological*, Zone Books, 1991.

Coleridge, Samuel Taylor, *Biographia Literaria*, 1817.

Cooper, John M., and Procopé, J. F. (eds), 'On Anger', in *Seneca: Moral and Political Essays*, Cambridge University Press, 1995.

Crystal, Ben, *Shakespeare on Toast*, Icon Books, 2009.

Dabhoiwala, F., *The Origins of Sex: A History of the First Sexual Revolution*, Penguin, 2013.

Damasio, A. R., *Descartes' Error: Emotion, Reason and the Human Brain*, Penguin, 2005.

Damasio, A. R., *The Feeling of What Happens*, Harcourt Brace & Co., 2000.

Darwin, C., *The Expression of the Emotions in Man and Animals*, John Murray, 1872.

Davidson, R. J. (and Begley, S.), *The Emotional Life of Your Brain*, Hudson Street Press (Penguin), 2012.

Diderot, D., *The Paradox of Acting* (trans. Walter H. Pollock), Chatto and Windus, 1883.

Doidge, Norman, *The Brain That Changes Itself*, Penguin, 2007.

Drury, M. O'C., 'Conversations with Wittgenstein', in Rhees, R. (ed.), *Ludwig Wittgenstein: Personal Recollections*, Rowman and Littlefield, 1981.

Dumit, J., *Picturing Personhood: Brain Scans and Biomedical Identity*, Princeton University Press, 2003.

Eagleman, D., *Incognito: The Secret Lives of the Brain*, Pantheon Books, 2011.

Eisenberger, N. I., and Lieberman, M. D., 'Why it hurts to be left out. The neurocognitive overlap between physical and social pain', in Williams, K. D., Forgas, J. P., and von Hippel, W. (eds), *The Social Outcast: Ostracism, Social Exclusion, Rejection, and Bullying*, Cambridge University Press, 2005.

Ekman, P., *Emotions Revealed*, Henry Holt and Company, 2003.

Fisher, H., *Why Him, Why Her*, One World Publications, 2009.

Fitzpatrick, S., 'Functional brain imaging: Neuro-turn or wrong turn?', in Littlefield, M., and Johnson, J. M. (eds), *The Neuroscientific Turn: Transdisciplinarity in the Age of the Brain*, University of Michigan Press, 2012.

Frances, Allen, *Saving Normal: An Insider Revolts against Out-of-control Psychiatric Diagnosis, DSM-5, Big Pharma, and the Medicalization of Everyday Life*, William Morrow, 2013.

Freshwater, H., *Theatre and the Audience*, Palgrave Macmillan, 2009.

Freud, S., *Civilization and its Discontents* (originally published 1930), Penguin, 2002.

Freud, S., *The Interpretation of Dreams* (originally published by Macmillan Company, 1913), Forgotten Books, 2012.

Freud, S., *Introductory Lectures on Psycho-analysis (Part III)*, Vol. XVI (1917), The Standard Edition of the Complete Psychological Works of Sigmund Freud. Lecture XXV: Anxiety.

Freud, S., *Mourning and Melancholia* (originally published 1917), 14th edn, Vintage, 1998.

Freud, S., *New Introductory Lectures on Psychoanalysis*, W. W. Norton and Company, 1933.

Freud, S., *Studies on Hysteria* (trans. J. Stratchey) (originally published 1895), Basic Books, 1957.

Ginsberg, A., *Howl, Kaddish and Other Poems*, Penguin Classics, 2009.

Goldstein, T., 'Responses to and judgments of acting on film', in Kaufman, J. C., and Simonton, D. K. (eds), *The Social Science of Cinema*, Oxford University Press, 2012.

Graham-Dixon, Andrew, *Caravaggio: A Life Sacred and Profane*, Penguin, 2010.

Gray, P., 'From Russia to America: A critical chronology', in Munk, E., (ed.), *Stanislavsky and America*, Hill and Wang, 1966.

Gruenberg, A. M., Goldstein, R. D., and Pincus, H. A., 'Classification of depression: Research and diagnostic criteria: DSM-IV and ICD-10', in Licinio, J., and Wong, M. L. (eds), *Biology of Depression: From Novel Insights to Therapeutic Strategies*, Wiley-VCH Verlag, 2005.

Hauser, M. D., *Moral Minds: How Nature Designed Our Universal Sense of Right and Wrong*, Ecco/HarperCollins, 2006.

Healy, D., *The Anti-depressant Era*, Harvard University Press, 1997.

Heidegger, M., *Being and Time* (originally published 1927), Harper & Row, 1962.

Hundert, E. J., *The Enlightenment's Fable: Bernard Mandeville and the Discovery of Society*, Cambridge University Press, 1994.

Illouz, Eva, *Cold Intimacies*, Polity, 2007.

Jackson, S. W., *Melancholia and Depression: From Hippocratic Times to Modern Times*, Yale University Press, 1986.

Kahneman, D., *Thinking Fast and Slow*, Farrar, Straus & Giroux, 2011.

Kandel, E., *The Age of Insight*, Random House, 2012.

Kübler-Ross, E., and Kessler, D., *On Grief and Grieving*, Scribner, 2007.

Larkin, P., *The Complete Poems*, Faber and Faber, 2012.

Layard, R., *Happiness: Lessons from a New Science*, Penguin, 2005.

Lipps, T., *Grundlegung der Aesthetik*, Engelmann, 1903.

Lloyd, G. E. R. (ed.), *Hippocratic Writings* (trans. J. Chadwyck and W. N. Mann), Penguin, 1978.

McLean, Paul, *The Triune Brain*, Plenum Press, 1990.

Metzl, Jonathan Michel, *Prozac on the Couch: Prescribing Gender in the Era of Wonder Drugs*, Duke University Press, 2003.

Miller, E. K., and Buschman, T. J., 'Rules through recursion: How interactions between the frontal cortex and basal ganglia may build abstract, complex rules from concrete, simple ones', in Bunge, S., and Wallis, J. (eds), *Neuroscience of Rule-Guided Behavior*, Oxford University Press, 2007.

Panksepp, J., *Affective Neuroscience*, Oxford University Press, 1998.

Prodger, P., *Darwin's Camera: Art and Photography in the Theory of Evolution*, Oxford University Press, 2009.

Prose, Francine, *Caravaggio, Painter of Miracles*, Harper Perennial, 2010.

Provine, R. R., *Curious Behavior: Yawning, Laughing, Hiccupping and Beyond*, Belknap Press (Harvard University Press), 2012.

Provine, R., *Laughter: A Scientific Investigation*, Viking, 2000.

Ramachandran, V. S., *The Tell-Tale Brain*, William Heinemann, Random House, 2011.

Ramon y Cajal, S., *Advice for a Young Investigator*, MIT Press, 1999.
Reeve, C. D. C., *Plato on Love*, Hackett Publishing Company, 2006.
Rilke, R. M., *Letters to a Young Poet*, W. W. Norton, 1993.
Rogers, James, *Dictionary of Clichés*, Wing Books, 1970.
Rose, S., *Lifelines: Life Beyond the Gene*, Oxford University Press, 2003.
Rose, S., and Rose, H., *Alas Poor Darwin: Arguments against Evolutionary Psychology*, Random House, 2000.
Sacks, O., *The Mind's Eye*, Picador, 2010.
Salecl, R., *On Anxiety*, Routledge, 2004.
Sawoski, P., *The Stanislavsky System: Growth and Methodology*, Teaching Material, Santa Monica College, Spring 2010.
Shaw, G. B., *Our Theatres in the Nineties*, Vol. I of *Collected Works*, Constable and Company, 1932.
Smith, M., *Small Comfort: A History of the Minor Tranquilizers*, Praeger, 1985.
Stanislavsky, C., *An Actor Prepares*, Theatre Art Books, 1936 (1983 printing).
Stanislavsky, C., *Creating a Role*, Methuen, 1961.
Stendhal, *On Love* (originally published 1822), Penguin, 1975, 2004.
Stewart, D. E., *The Arrow of Love*, Associated University Presses, 2010.
Tangney, June Price, and Dearing, Ronda L., *Shame and Guilt*, Guilford Press, 2000.
Titchener, E. B., *Lectures on the Experimental Psychology of Thought Processes*, Macmillan, 1909.
Tone, Andrea, *The Age of Anxiety: A History of America's Turbulent Affair with Tranquilizers*, Basic Books, 2008.
Unger, P., *Living High and Letting Die: Our Illusion of Innocence*, Oxford University Press, 1996.
Vischer, R., *Über das optische Formgefühl: Ein Beiträg zur Ästhetik*, Credner, 1873.
Weber, M. (ed. D. Owen and T. B. Strong), *The Vocation Lectures*, Hackett Publishing Company, 2004.
Wilde, O., *The Picture of Dorian Gray*, Penguin, ebook, 2006.
Wilson, E. O. (ed.), *From So Simple a Beginning: The Four Great Books of Charles Darwin*, Norton, 2006.
Wittgenstein, L., *Lectures on Aesthetics* (originally published 1966), University of California Press, 40th Anniversary Edition, 2007.
Wittgenstein, L., *Philosophical Investigations* (trans. G. E. M. Anscombe), Oxford University Press, 1953.
Wittgenstein, L., *Remarks on the Philosophy of Psychology*, Vol. II (trans. G. E. M. Anscombe), University of Chicago Press, 1980.

**Internet**

American Psychiatric Association's proposals and preliminary drafts for the fifth edition of the Diagnostic and Statistical Manual of Mental Disorders: www.dsm5.org

Award of Nobel Prize in 1970 to Sir Bernard Katz, Ulf von Euler and Julius Axelrod for their discovery of serotonin re-uptake: http://www.nobelprize.org/nobel_prizes/medicine/laureates/1970/

Ayers, Robert: interview with Marina Abramovic: http://www. askyfilledwithshootingstars.com/wordpress/?p=1197

Brooks, David: interview with Antonio Damasio on FORA.tv: http://fora.tv/ 2009/07/04/Antonio_Damasio_This_Time_With_Feeling

Fallon, Dr Jim: talks about his story in his own TED Talk: http://www.youtube.com/watch?v=u2V0vOFexY4

Frances, Allen: warning against the creation of a new psychiatric category for grief: http://www.psychologytoday.com/blog/dsm5-in-distress/201008/good-grief-vs-major-depressive-disorder

Frazzetto, G., *et al.*, *Never Mind*, premiered at the Sophiensäle, Berlin, on 25 January 2012: http://www.sophiensaele.com/produktionen.php?IDstueck=901

'From describing to nudging: Choice of transportation after a terrorist attack in London', a study of the impact of the July bombings on Londoners' travel behaviour: http://research.create.usc.edu/project_summaries/67

Greene, Joshua: speech at a conference on the science of morality: http://www.edge.org/3rd_culture/morality10/morality.greene.html

Hagan, C., 'Geneticists studying Connecticut shooter's DNA', CNN online, 28 December 2012.

Ignobel Prizes: www.improbable.com

Insel, Thomas, 'Transforming Diagnosis': http://www.nimh.nih.gov/about/director/directors-biography.shtml

LeDoux, Joseph: interview in *Slate* online magazine: http://www.slate.com/articles/life/brains/2007/04/brain_lessons.html

MacKinnon, Eli, 'Eduardo Leite dubbed modern-day Phineas Gage after pole pierces his brain', HuffPost Science, 22 August 2012.

NHS fact sheet on generalized anxiety disorder: http://www.nhs.uk/conditions/anxiety/Pages/Introduction.aspx

Pies, R., 'Why psychiatry needs to scrap the DSM system: An immodest proposal': http://psychcentral.com/blog/archives/2012/01/07/why-psychiatry-needs-to-scrap-the-dsm-system-an-immodest-proposal/

Statistics page for anxiety disorders of the National Institute of Mental Health: http://www.nimh.nih.gov/statistics/1ANYANX_ADULT.shtml

**Printed press**

'After shock', *Guardian*, 17 June 2006.

Baldwin, J., Letter: 'From a region in my mind', *New Yorker*, 17 November 1962.

Barron, J., 'Nation reels after gunman massacres 20 children at school in Connecticut', *New York Times*, 14 December 2012.

Campbell, D., 'Recession causes surge in mental health problems', *Guardian*, 1 April 2010.

Dyson, F., 'How to dispel your illusions', *New York Review of Books*, 22 December 2011.
Gautam, N., 'What's on Jim Fallon's Mind?', *Wall Street Journal*, 30 November 2009.
Gopnik, Adam, 'One more massacre', *New Yorker*, 20 July 2012.
Helm, Toby, 'Victims of recession to get free therapy', *Guardian*, 8 March 2009.
Hirschberg, Lynn: interview with Daniel Day-Lewis, *Daily Telegraph*, 8 December 2007.
Johnson, Carolyn Y., 'Ex-Harvard scientist fabricated, manipulated data, report says', *Boston Globe*, 5 September 2012.
Lee, Jennifer, 'Is that an emoticon in 1862?', *New York Times*, 19 January 2009.
McGrath, Charles: interview with Daniel Day-Lewis, *New York Times*, 31 October 2012.
'Medicine: To Nirvana with Miltown', *Time*, 7 July 1958.
Neville, S., 'GlaxoSmithKline fined $3bn after bribing doctors to increase drug sales', *Guardian*, 3 July 2012.
Pinsky, Robert, 'Grief', *New York Review of Books*, 7 June 2012.
Report on massacre carried out by James Holmes at Aurora, Colorado, USA, on 19 July 2012, Associated Press, 31 August 2012.
Sample, Ian, 'Female orgasm captured in series of brain scans', *Guardian*, 14 November 2011.
Smith, Z., 'Joy', *New York Review of Books*, 10 January 2013.
Tomlinson, Simon, *Daily Mail*, 11 December 2012.
Total number of deaths registered in 2011 in England and Wales as reported by the Office for National Statistics was 484,367: *Guardian*, 6 November 2012.
Widdicombe, Lizzie, 'Shots', *New Yorker*, 3 September 2012.

### Scientific journals

Abel, E. L., and Kruger, M. L., 'Smile intensity in photographs predicts longevity', *Psychological Science*, 21 (2010).
Abraham, A., and von Cramon, D. Y., 'Reality-relevance? Insights from spontaneous modulations of the brain's default network when telling apart reality from fiction', *PLOS One*, 4 (2009).
Abraham, A., von Cramon, D. Y., and Schubotz, R. I., 'Meeting George Bush versus Cinderella: The neural response when telling apart what is real from what is fictional in the context of our reality', *Journal of Cognitive Neuroscience*, 20 (2008).
Adolphs, R., Tranel, D., Damasio, H., and Damasio, A., 'Impaired recognition of emotion in facial expressions following bilateral damage to the human amygdala', *Nature*, 372 (1995).
Amodio, D. M., and Frith, C. D., 'Meeting of minds: The medial frontal cortex and social cognition', *Nature Reviews Neuroscience*, 7 (2006).

Amorapanth, P., LeDoux, J. E., and Nader, K., 'Different lateral amygdala outputs mediate reactions and actions elicited by a fear-arousing stimulus', *Nature Reviews Neuroscience*, 3 (2000).
Anderson, S. W., Bechara, A., Damasio, H., Tranel, D., and Damasio, A. R., 'Impairment of social and moral behavior related to early damage in human prefrontal cortex', *Nature Reviews Neuroscience*, 2 (1999).
Aron, A., Fisher, H., Mashek, D. J., Strong, G., Li, H., and Brown, L. L., 'Reward, motivation and emotion systems associated with early stage intense romantic love', *Journal of Neurophysiology*, 94 (2005).
Aspinwall, L. G., Brown, T. R., and Tabery, J., 'The double-edged sword: Does biomechanism increase or decrease judges' sentencing of psychopaths?', *Science*, 337 (2012).
Bartels, A., and Zeki, S., 'The neural basis of romantic love', *Neuroreport*, 11 (2000).
Bartels, A., and Zeki, S., 'The neural correlates of maternal and romantic love', *Neuroimage*, 21 (2004).
Basile, B., Mancini, F., Macaluso, E., Caltagirone, C., Frackowiak, R. S., and Bozzali, M., 'Deontological and altruistic guilt: Evidence for distinct neurobiological substrates', *Human Brain Mapping*, 32 (2011).
Bechara, A., Damasio, H., Tranel, D., and Damasio, A. R., 'Deciding advantageously before knowing the advantageous strategy', *Science*, 275 (1997).
Bennett, C. M., Baird, A. A., Miller, M. B., and Wolford, G. L., 'Neural correlates of interspecies perspective taking in the post-mortem Atlantic salmon: An argument for proper multiple comparisons correction', *Journal of Serendipitous and Unexpected Results*, 1 (2010).
Berridge, K. C., '"Liking" and "wanting" food rewards: Brain substrates and roles in eating disorders', *Physiology and Behavior*, 97 (2009).
Beutel, M. E., Stark, R., Pan, H., Silbersweig, D., and Dietrich, S., 'Changes of brain activation pre-post short-term psychodynamic inpatient psychotherapy: An fMRI study of panic disorder patients', *Psychiatry Research*, 184 (2010).
Blair, R. J. R., and Cipolotti, L., 'Impaired social response reversal. A case of "acquired sociopathy"', *Brain*, 123 (2000).
Blood, A. J., and Zatorre, R. J., 'Intensely pleasurable responses to music correlate with activity in brain regions implicated in reward and emotion', *Proceedings of the National Academy of Sciences*, 98 (2001).
'Brain Waves 4: Neuroscience and the law', Royal Society, London, December 2011.
Brunner, H. G., *et al.*, 'Abnormal behaviour associated with a point mutation in the structural gene for monoamine oxidase A', *Science*, 262 (1993).
Brunner, H. G., *et al.*, 'X-linked borderline mental retardation with prominent behavioural disturbance: Phenotype, genetic localization, and evidence for disturbed monoamine metabolism', *American Journal of Human Genetics*, 52 (1993).

Burgdorf, J., Panksepp, J., Brudzynski, S. M., Kroes, R., and Moskal, J. R., 'Breeding for 50-kHz positive affective vocalization in rats', *Behavior Genetics*, 35 (2005).

Bylsma, L. M., Vingerhoets, A. J. J. M., and Rottenberg, J., 'When is crying cathartic? An international study', *Journal of Social and Clinical Psychology*, 27 (2008).

Campbell, R., Heywood, C., Cowey, A., Regard, M., and Landis, T., 'Sensitivity to eye gaze in prosopagnosic patients and monkeys with superior temporal sulcus ablation', *Neuropsychologia*, 28 (1990).

Capgras, J., 'L'illusion des "sosies" dans un délire systématisé chronique', *Bulletin de la Société Clinique de Médecine Mentale*, 11 (1923).

Carola, V., Frazzetto, G., and Gross, C., 'Identifying interactions between genes and early environment in the mouse', *Genes, Brain and Behavior*, 5 (2006).

Carr, L., Iacoboni, M., Dubeau, M. C., Mazziotta, J. C., and Lenzi, G. L., 'Neural mechanisms of empathy in humans: A relay from neural systems for imitation to limbic areas', *Proceedings of the National Academy of Sciences*, 100 (2003).

Cases, O., Seif, I., Grimsby, J., *et al.*, 'Aggressive behavior and altered amounts of brain serotonin and norepinephrine in mice lacking MAOA', *Science*, 268 (1995).

Caspi, A., *et al.*, 'Role of genotype in the cycle of violence in maltreated children', *Science*, 297 (2002).

Chakrabarti, B., Bullmore, E., and Baron-Cohen, S., 'Empathising with basic emotions: Common and discrete neural substrates', *Social Neuroscience*, 1 (2006).

Chang, P. P., Ford, D. E., Meoni, L. A., *et al.*, 'Anger in young men and subsequent premature cardiovascular disease', *Archives of Internal Medicine*, 162 (2002).

Cohen, S., Alper, C. M., Doyle, W. J., *et al.*, 'Positive emotional style predicts resistance to illness after experimental exposure to rhinovirus or influenza A virus', *Psychosomatic Medicine*, 68 (2006).

Collier, R., 'Recession stresses mental health', *Canadian Medical Association Journal*, 181 (2009).

Damasio, H., *et al.*, 'The return of Phineas Gage: Clues about the brain from the skull of a famous patient', *Science*, 264 (1994).

Davidson, R. J., Ekman, P., Saron, C. D., Senulis, J. A., and Friesen, W. V., 'Approach-Withdrawal and cerebral asymmetry: Emotional expression and brain physiology I', *Journal of Personality and Social Psychology*, 58 (1990).

Davidson, R. J., and Fox, N. A. 'Asymmetrical brain activity discriminates between positive versus negative affective stimuli in human infants', *Science*, 218 (1982).

Davidson, R. J., and Fox, N. A., 'Frontal brain asymmetry predicts infants' response to maternal separation', *Journal of Abnormal Psychology*, 98 (1989).

Davidson, R. J., Putnam, K. M., and Larson, C. L., 'Dysfunction in the neural circuitry of emotion regulation – a possible prelude to violence', *Science*, 289 (2000).

Debruille, J. B., and Stip, E., 'Capgras syndrome: Evolution of the hypothesis', *Canadian Journal of Psychiatry*, 41 (1996).

Diener, E., and Biswar-Diener, R., 'Will money increase subjective well-being?', *Social Indicators Research*, 57 (2002).

Dunbar, R. I. M., Baron, R., Frangou, A., *et al.*, 'Social laughter is correlated with an elevated pain threshold', *Proceedings of the Royal Society B: Biological Sciences*, 279 (2011).

Dunn, E. W., Aknin, L. B., and Norton, M. I., 'Spending money on others promotes happiness', *Science*, 319 (2008).

Eisenberg, N., 'Emotion, regulation and moral development', *Annual Review of Psychology*, 51 (2000).

Eisenberger, N. I., Lieberman, M. D., and Williams, K. D., 'Does rejection hurt? An fMRI study of social exclusion', *Science*, 302 (2003).

Ekman, P., Davidson, R. J., and Friesen, W. V., 'The Duchenne smile: Emotional expression and brain physiology II', *Journal of Personality and Social Psychology*, 58 (1990).

Escobedo, J. R., and Adolphs, R., 'Becoming a better person: Temporal remoteness biases autobiographical memories for moral events', *Emotion*, 10 (2010).

Fakra, E., Hyde, L. W., Gorka, A., Fisher, P. M., Munoz, K. E., Kimak, M., Halder, I., Ferrell, R. E., Manuck, S. B., and Hariri, A. R., 'Effects of Htr1a C(-1019) G on amygdala reactivity and trait anxiety', *Archives of General Psychiatry*, 66 (2009).

Fendez, M. F., 'The neurobiology of moral behavior: Review and neuropsychiatric implications', *CNS Spectre*, 14 (2009).

Feresin, E., 'Lighter sentence for murderer with "bad genes"', *Nature*, 30 October 2009.

Finkel, E. J., Eastwick, P. W., Karney, B. R., Reis, H. T., and Sprecher, S., 'Online dating: A critical analysis from the perspective of psychological science', *Psychological Science in the Public Interest*, 13 (2012).

Fiorino, F., Coury, A., and Phillips, A. G., 'Dynamic changes in nucleus accumbens dopamine efflux during the Coolidge effect in male rats', *Journal of Neuroscience*, 17 (1997).

Foot, P., 'A new definition' (Recipes for Happiness), *British Medical Journal*, 321 (2000).

Frazzetto, G., 'Genetics of behavior and psychiatric disorders: From the laboratory to society and back', *Current Science*, 97 (2009).

Frazzetto, G.,'Powerful Acts', *Nature*, 482 (2012).

Frazzetto, G., *et al.*, 'Early trauma and increased risk for physical aggression during adulthood: The moderating role of MAOA genotype', *PLOS One*, 5, Issue 2 (2007).

Freedberg, D., and Gallese, V., 'Motion, emotion and empathy in esthetic experience', *Trends in Cognitive Sciences*, 11 (2007).

Frydman, C., Camerer, C., Bossaerts, P., and Rangel, A., 'MAOA-L carriers are better at making optimal financial decisions under risk', *Proceedings of the Royal Society*, 278 (2010).

Gainotti, G., 'Emotional behavior and hemispheric side of the lesion', *Cortex*, 8 (1972).

Gallese, V., 'The roots of empathy: The shared manifold hypothesis and the neural basis of intersubjectivity', *Psychopathology*, 36 (2003).

Gallese, V., Fadiga, L., and Rizzolatti, G., 'Action recognition in the premotor cortex', *Brain*, 119 (1996).

Goldstein, R., 'Thrills in response to music and other stimuli', *Physiological Psychology*, 8 (1980).

Gozzi, A., Jain, A., Giovanelli, A., *et al.*, 'A neural switch for active and passive fear', *Neuron*, 67 (2010).

Greene, J., 'From neural "is" to moral "ought": What are the moral implications of neuroscientific moral psychology?', *Nature Reviews Neuroscience*, 4 (2003).

Greene, J. D., Sommerville, R. B., Nystrom, L. E., Darley, J. M., and Cohen, J. D., 'An fMRI investigation of emotional engagement in moral judgment', *Science*, 293 (2001).

Haker, H., Kawohl, W., Herwig, U., and Rössler, W., 'Mirror neuron activity during contagious yawning – an fMRI study', *Brain Imaging and Behavior* (7 July 2012: electronic abstract ahead of publication).

Hamann, S., 'Mapping discrete and dimensional emotions onto the brain: Controversies and consensus', *Trends in Cognitive Sciences*, 16 (2012).

Hammer, M., 'An identified neuron mediates the unconditioned stimulus in associative olfactory learning in honeybees', *Nature*, 366 (1993).

Hammer, M., and Menzel, R., 'Learning and memory in the honeybee', *Journal of Neuroscience*, 15 (1995).

Harker, L. A., and Keltner, D., 'Expressions of positive emotion in women's college yearbook pictures and their relationship to personality and life outcomes across adulthood', *Journal of Personality and Social Psychology*, 80 (2001).

Harlow, J., *Publications of the Massachusetts Medical Society*, 2 (1868).

Harris, J. C., 'Caravaggio's Narcissus', *American Journal of Psychiatry*, 67 (2010).

Hasson, O., 'Emotional Tears as Biological Signals', *Evolutionary Psychology*, 7 (2009).

Hatfield, E., and Sprecher, S., 'Measuring passionate love in an intimate relation', *Journal of Adolescence*, 9 (1986).

Hazan, C., and Shaver, P., 'Romantic love conceptualized as an attachment process', *Journal of Personality and Social Psychology*, 52 (1987).

Hirstein, W., and Ramachandran, V. S., 'Capgras syndrome: A novel probe for understanding the neural representation of the identity and familiarity of persons', *Proceedings of the Royal Society B: Biological Sciences*, 264 (1997).

Holstege, G., Georgiadis, J. R., Paans, A. M. J., *et al.*, 'Brain activation during human male ejaculation', *Journal of Neuroscience*, 23 (2003).

Holt-Lunstad, J., Smith, T. B., and Layton, J. B., 'Social relationships and mortality risk: A meta-analytic review', *PLOS Medicine*, 7, Issue 7 (2010).

Iacoboni, M., *et al.*, 'Cortical mechanisms of human imitation', *Science*, 286 (2003).

Insel, T. R., 'The challenge of translation in social neuroscience: A review of oxytocin, vasopressin, and affiliative behavior', *Neuron*, 65 (2010).

Insel, T. R., 'Next-generation treatments for mental disorders', *Science Translational Medicine*, 4, Issue 155 (2012).

Insel, T., 'Oxytocin – a neuropeptide for affiliation: Evidence from behavioural, receptor autoradiographic, and comparative studies', *Psychoneuroendocrinology*, 17 (1992).

Isen, A., Daubman, K. A., and Nowicki, G. P., 'Positive affect facilitates creative problem solving', *Journal of Personality and Social Psychology*, 52 (1987).

Isen, A. M., Johnson, M. M. S., Mertz, E., and Robinson, G. F., 'The influence of positive affect on the unusualness of word associations', *Journal of Personality and Social Psychology*, 48 (1985).

Jabbi, M., and Keysers, C., 'Inferior frontal gyrus activity triggers anterior insula response to emotional facial expressions', *Emotion*, 8 (2008).

James, W., 'What is an emotion?', *Mind*, 9 (1884).

Jones, D., 'The depths of disgust', *Nature*, 447 (2007).

Jueptner, M., and Weiller, C., 'Review: Does measurement of regional cerebral blood flow reflect synaptic activity? Implications for PET and fMRI', *Neuroimage*, 2 (1995).

Kampe, K., Frith, C. D., Dolan, R. J., and Frith, U., 'Reward value of attractiveness and gaze', *Nature*, 413 (2001).

Kedia, G., Berthoz, S., Wessa, M., Hilton, D., and Martinot, J. L., 'An agent harms a victim: A functional magnetic resonance imaging study on specific moral emotions', *Journal of Cognitive Neuroscience*, 20 (2008).

Keltner, D., and Bonanno, G. A., 'A study of laughter and dissociation: Distinct correlates of laughter and smiling during bereavement', *Journal of Personality and Social Psychology*, 4 (1997).

Kendler, K. S., Myers, J. M. S., and Zisook, S., 'Does bereavement-related major depression differ from major depression associated with other stressful life events?', *American Journal of Psychiatry*, 165 (2008).

Kessler, R. C., Chiu, W. T., Demler, O., *et al.*, 'Prevalence, severity, and comorbidity of twelve-month DSM-IV disorders in the National Comorbidity Survey Replication (NCS-R)', *Archives of General Psychiatry*, 62 (2005).

Keysers, C., Wicker, B., Gazzola, V., Anton, J. L., Fogassi, L., and Gallese, V., 'A touching sight: SII/PV activation during the observation and experience of touch', *Neuron*, 42 (2004).

Kim-Cohen, J., *et al.*, 'MAOA, maltreatment, and gene–environment interaction predicting children's mental health: New evidence and a meta-analysis', *Molecular Psychiatry*, 11 (2006).

Kirsch, I., Deacon, B. F., Huedo-Medina, T. B., *et al.*, 'Initial severity and antidepressant benefits: A meta-analysis of data submitted to the Food and Drug Administration', *PLOS Medicine*, 5, Issue 2 (2008).

Klein, D. F., 'Delineation of two drug responsive anxiety syndromes', *Psychopharmacologia*, 5 (1964).

Kleinman, Arthur, editorial in *Lancet*, 379, 18 February 2012.

Kok, B. E., Coffey, E. A., Cohn, M. A., *et al.*, 'How positive emotions build physical health: Perceived positive social connections account for the upward spiral between positive emotions and vagal tone', *Psychological Science*, in press.

Kok, B. E., and Fredrickson, B. L., 'Upward spirals of the heart: Autonomic flexibility, as indexed by vagal tone, reciprocally and prospectively predicts positive emotions and social connectedness', *Biological Psychology*, 85 (2010).

Komisaruk, B. R., and Whipple, B., 'Functional MRI of the brain during orgasm in women', *Annual Review of Sex Research*, 16 (2005).

Komisaruk, B. R., Whipple, B., Crawford, A., *et al.*, 'Brain activation during vaginocervical self-stimulation and orgasm in women with complete spinal cord injury: fMRI evidence of mediation by the Vagus nerves', *Brain Research*, 1024 (2004).

Lacasse, J. R., and Leo, J., 'Serotonin and depression: A disconnect between the advertisements and the scientific literature', *PLOS Medicine*, 2, Issue 12 (2005).

LeDoux, Joseph, 'Rethinking the emotional brain', *Neuron*, 73 (2012).

LeDoux, J., and Gorman, J. M., 'A call to action: Overcoming anxiety through active coping', *American Journal of Psychiatry*, 158 (2001).

Lee, S. W. S., and Scharz, N., 'Washing away post-decisional dissonance', *Science*, 328 (2010).

Lennard, H. L., Epstein, L. J., Bernstein, A., and Ranson, D. C., 'Hazards implicit in prescribing psychoactive drugs', *Science*, 169 (1970).

Lenzer, J., 'Bush plans to screen whole US population for mental illness', *British Medical Journal*, 328 (2004).

'Living with grief', editorial, *Lancet*, 379 (2011).

Logothetis, N. K., 'What we can do and what we cannot do with fMRI', *Nature*, 453 (2008).

Maher, B., 'Poll results: Look who's doping', *Nature*, 452 (2008).

Mar, R. A., and Oatley, K., 'The function of fiction is the abstraction and simulation of social experience', *Perspectives on Psychological Science*, 3 (2008).

Marazziti, D., and Canale, D., 'Hormonal changes when falling in love', *Psychoneuroendocrinology*, 29 (2004).

Mascolo, M. F., 'Wittgenstein and the discursive analysis of emotion', *New Ideas in Psychology*, 27 (2009).

McCabe, D. P., and Castel, A. D., 'Seeing is believing: The effect of brain images on judgments of scientific reasoning', *Cognition*, 107 (2008).

McGoey, L., 'On the will to ignorance in bureaucracy', *Economy and Society*, 36 (2007).

McGoey, L., 'Profitable failure: Antidepressant drugs and the triumph of flawed experiments', *History of the Human Sciences*, 23 (2010).

McRae, R. R., and Costa, P. T., 'Validation of the five-factor model across instruments and observers', *Journal of Personality and Social Psychology*, 52 (1987).

Metz-Lutz, M. N., Bressan, Y., Heider, N., and Otzenberger, H., 'What physiological and cerebral traces tell us about adhesion to fiction during theater-watching', *Frontiers in Human Neuroscience*, 4 (2010).

Meyer-Lindenberg, A., Buckholtz, J. W., Kolachana, B., *et al.*, 'Neural mechanisms of genetic risk for impulsivity and violence in humans', *Proceedings of the National Academy of Sciences*, 103 (2006).

Miller, G., 'Is Pharma running out of brainy ideas?', *Science*, 329 (2010).

Moll, J., *et al.*, 'Human fronto-mesolimbic networks guide decisions about charitable donation', *Proceedings of the National Academy of Sciences*, 103 (2006).

Moll, J., Oliveira-Souza, R., Garrido, G. J., Bramati, I. E., Caparelli-Daquer, E. M., Paiva, M., Zahn, R., and Grafman, J., 'The Self as a moral agent: Linking the neural bases of social agency and moral sensitivity', *Social Neuroscience*, 2 (2007).

Montague, P. R., Dayan, P., Person, C., and Sejnowski, T. J., 'Bee foraging in uncertain environments using predictive Hebbian learning', *Nature*, 377 (1995).

Nesse, R., 'Proximate and evolutionary studies of anxiety, stress and depression: Synergy at the interface', *Neuroscience and Biobehavioral Reviews*, 23 (1999).

O'Connor, M. F., Wellisch, D. K., Stanton, A. L., Eisenberger, E. I., Irwin, M. R., and Lieberman, M. D., 'Craving love? Enduring grief activates brain's reward center', *Neuroimage*, 42 (2008).

Olds, J., and Milner, P., 'Positive reinforcement produced by electrical stimulation of septal area and other regions of rat brain', *Journal of Comparative and Physiological Psychology*, 47 (1954).

Pauling, L., and Coryell, C., 'The magnetic properties and structure of hemoglobin', *Proceedings of the National Academy of Sciences*, 22 (1936).

Peterson, C., Nansook, P., and Seligman, M. E. P., 'Orientations to happiness and life satisfaction: The full life versus the empty life', *Journal of Happiness Studies*, 6 (2005).

Pletscher, A., Shore, P. A., and Brodie, B. B., 'Serotonin release as a possible mechanism of reserpine action', *Science*, 122 (1955).

Porges, S. W., 'The polyvagal perspective', *Biological Psychology*, 74 (2007).
Porter, R., 'Happy hedonists' (Recipes for Happiness), *British Medical Journal*, 321 (2000).
Prigerson, H. G., Horowitz, M. J., Jacobs, S. C., *et al.*, 'Prolonged Grief Disorder: Psychometric validation of criteria proposed for DSM-V and ICD-11', *PLOS Medicine*, 6, Issue 8 (2009).
Provine, R. R., 'Emotional tears and NGF: A biographical appreciation and research beginning', *Archives Italiennes de Biologie*, 149 (2011).
Provine, R., 'Laughter', *American Scientist*, 84 (1996).
Provine, R., and Fischer, K. R., 'Laughing, smiling and talking: Relation to sleeping and social context in humans', *Ethology*, 83 (1989).
Provine, R. R., Krosnowski, K. A., and Brocato, N. W., 'Tearing: Breakthrough in human emotional signaling', *Evolutionary Psychology*, 7 (2009).
Raine, A., Buchsbaum, M., and LaCasse, L., 'Brain abnormalities in murderers indicated by positron emission tomography', *Biological Psychiatry*, 42 (1997).
Raine, A., Meloy, J. R., Bihrle, S., Stoddard, J., LaCasse, L., and Buchsbaum, M. S., 'Reduced prefrontal and increased subcortical brain functioning assessed using positron emission tomography in predatory and affective murderers', *Behavioural Sciences and the Law*, 16 (1998).
Rakersting, A., Kroker, K., Horstmann, J., *et al.*, 'Association of MAO-A variant with complicated grief in major depression', *Neuropsychobiology*, 56 (2008).
Rakic, P., 'Evolution of the neocortex: Perspective from developmental biology', *Nature Reviews Neuroscience*, 10 (2010).
Rapp, A., Leube, D. T., Erb, M., Grodd, W., and Kircher, T. T., 'Neural correlates of metaphor processing', *Cognitive Brain Research*, 20 (2004).
Rizzolatti, G., and Craighero, L., 'The mirror-neuron system', *Annual Review of Neuroscience*, 27 (2004).
Rizzolatti, G., Fadiga, L., Gallese, V. and Fogassi, L., 'Premotor cortex and the recognition of motor actions', *Cognitive Brain Research*, 3 (1996).
Rose, N., 'Life, reason and history: Reading Georges Canguilhem today', *Economy and Society*, 27 (1998).
Rose, N., 'Neurochemical selves', *Society*, 41 (2003).
Sabol, S., *et al.*, 'A functional polymorphism in the monoamine oxidase A gene promoter', *Human Genetics*, 103 (1998).
Sackeim, H. A., Greenberg, M. S., Weiman, A. L., *et al.*, 'Hemispheric asymmetry in the expression of positive and negative emotions', *Archives of Neurology*, 39 (1982).
Sahakian, B., and Morein-Zamir, S., 'Professor's little helper', *Nature*, 450 (2007).
Sauter, D. A., Eisner, F., Ekman, P., and Scott, S. K., 'Cross-cultural recognition of basic emotions through emotional vocalizations', *Proceedings of the National Academy of Sciences*, 107 (2010).

Schaffer, C. E., Davidson, R. J., and Saron, C., 'Frontal and parietal electroencephalogram asymmetry in depressed and nondepressed subjects', *Biological Psychiatry*, 18 (1987).

Schildkraut, J. J., 'The catecholamine hypothesis of affective disorders: A review of the supporting evidence', *American Journal of Psychiatry*, 122 (1965).

Schleim, S., 'Brains in context in the neurolaw debate: The examples of free will and "dangerous" brains', *International Journal for Law and Psychiatry*, 35 (2012).

Schultz, W., 'Multiple reward signals in the brain', *Nature Reviews Neuroscience*, 1 (2000).

Schultz, W., Apicella, P., and Ljungberg, T., 'Responses of monkey dopamine neurons to reward and conditioned stimuli during successive steps of learning a delayed response task', *Journal of Neuroscience*, 13 (1993).

Schultz, W., Apicella, P., Scarnati, E., and Ljungberg, T., 'Neuronal activity in monkey ventral striatum related to the expectation of reward', *Journal of Neuroscience*, 12 (1992).

Schwartz, G. E., Davidson, R. J., and Maer, F., 'Right hemisphere lateralization for emotion in the human brain: Interactions with cognition', *Science*, 190 (1975).

Senju, A., and Johnson, M. H., 'The eye contact effect: Mechanisms and development', *Trends in Cognitive Sciences*, 13 (2009).

Shamay-Tsoory, S. G., Tibi-Elhanamy, Y., and Aharon-Petrez, J., 'The green-eyed monster and malicious joy: The neuroanatomical bases of envy and gloating (Schadenfreude)', *Brain*, 130 (2007).

Sloboda, J. A., 'Music structure and emotional response: Some empirical findings', *Psychology of Music*, 19 (1991).

Sluzkin, C., 'On sorrow: Medical advice from Ishaq ben Sulayman al-Israeli, 1000 years ago', *American Journal of Psychiatry*, 167, 5 (2010).

Smith, C. U. M., 'The triune brain in antiquity: Plato, Aristotle, Erasistratus', *Journal of the History of the Neurosciences*, 19 (2010).

Smith, K., 'Trillion-dollar brain drain', *Nature*, 478 (2011).

Solms, M., 'Freud returns', *Scientific American*, May 2004.

Tafrate, R. C., Kassinove, H., and Dundin, L., 'Anger episodes in high- and low-trait anger community adults', *Journal of Clinical Psychology*, 58 (2002).

Takahashi, H., Kato, M., Matsuura, M., Mobbs, D., Suhara, T., and Okubo, Y., 'When your gain is my pain and your pain is my gain: Neural correlates of envy and Schadenfreude', *Science*, 323 (2009).

Takahashi, H., Yahata, N., Koeda, M., Matsuda, T., Asai, K., and Okubo, Y., 'Brain activation associated with evaluative processes of guilt and embarrassment: An fMRI study', *Neuroimage*, 23 (2004).

Toma, C. L., Hancock, J. T., and Ellison, N. B., 'Separating fact from fiction: An examination of deceptive self-representation on online dating profiles', *Personality and Social Psychology Bulletin*, 34 (2008).

Vul, E., Harris, C., Winkielman, P., and Pashler, H., 'Puzzlingly high correlations in fMRI studies of emotion, personality, and social cognition', *Perspectives on Psychological Science*, 4 (2009).

Wagner, U., N'Diaye, K., Ethofer, T., and Vuilleumier, P., 'Guilt-specific processing in the prefrontal cortex', *Cerebral Cortex*, 21 (2011).

Walum, H., Westberg, L., Henningsson, Jenae M., *et al.*, 'Genetic variation in the vasopressin receptor 1a gene (AVPR1A) associates with pair-bonding behavior in humans', *Proceedings of the National Academy of Sciences*, 105 (2008).

Warren, J. E., Sauter, D. A., Eisner, F., *et al.*, 'Positive emotions preferentially engage an auditory-motor "mirror" system', *Journal of Neuroscience*, 26 (2006).

Weng, H. Y., Fox, A. S., Shackman, A. J., Stodola, D. E., *et al.*, 'Compassion training alters altruism and neural responses to suffering', *Psychological Science*, in press 2013.

'Who calls the shots?', editorial after the massacre by James Holmes in Aurora, Colorado, USA, *Nature*, 488 (2012).

Wicker, B., Keysers, C., Plailly, J., Royet, J. P., Gallese, V., and Rizzolatti, G., 'Both of us disgusted in my insula: The common neural basis of seeing and feeling disgust', *Neuron*, 40 (2003).

Widom, C. S., and Brzustowicz, L. M., 'MAOA and the "cycle of violence": Childhood abuse and neglect, MAOA genotype, and risk for violent and antisocial behaviour', *Biological Psychiatry*, 60 (2006).

Wittchen, H. U., Jacobi, F., Rehm. A., Gustavsson, A., Svensson, M., Jönsson, B., Olesen, J., Allgulander, C., Alonso, J., Faravelli, C., Fratiglioni, L., Jennum, P., Lieb, R., Marcker, A., *et al.*, 'The size and burden of mental disorders and other disorders of the brain in Europe 2010', *European Neuropsychopharmacology*, 21 (2011).

Woolley, J. D., and Fields, H. L., 'Nucleus accumbens opioids regulate flavor-based preferences in food consumption', *Neuroscience*, 17 (2006).

Yang, Y., and Raine, A., 'Prefrontal structural and functional brain imaging findings in antisocial, violent, and psychopathic individuals: A meta-analysis', *Psychiatry Research*, 174 (2009).

Zeelenberg, M., and Breugelmans, S. M., 'The role of interpersonal harm in distinguishing regret from guilt', *Emotion*, 8 (2008).

Zeki, S., 'The neurobiology of love', *FEBS Letters*, 581 (2007).

Zeki, S., and Romaya, J. P., 'The brain reaction to viewing faces of opposite- and same-sex romantic partners', *PLOS One*, 5, Issue 12 (2010).

Zysset, S., Huber, O., Ferstl, E., and von Cramon, D. Y., 'The anterior frontomedian cortex and evaluative judgment: An fMRI study', *Neuroimage*, 15 (2002).

# Index

Note: Page numbers in *italics* refer to illustrations.